CONS
TE

LEVEL 1

R. L. FULLERTON
MSc, CEng, MIStructE, FIOB

OXFORD UNIVERSITY PRESS 1980

Oxford University Press, Walton Street, Oxford OX2 6DP

OXFORD LONDON GLASGOW
NEW YORK TORONTO MELBOURNE WELLINGTON
KUALA LUMPUR SINGAPORE JAKARTA HONG KONG TOKYO
DELHI BOMBAY CALCUTTA MADRAS KARACHI
NAIROBI DAR ES SALAAM CAPE TOWN

© R. Fullerton 1980

Published in the United States
by Oxford University Press, New York

All rights reserved. No part of this publication may be reproduced, stored in a retrieval system, or transmitted, in any form or by any means, electronic, mechanical, photocopying, recording, or otherwise, without the prior permission of Oxford University Press

British Library Cataloguing in Publication Data

Fullerton, Richard Lewis
   Construction technology, Level 1.
   1. Building
   I. Title
   690'.02'46   TH145   79-41320

ISBN 0-19-859519-0
ISBN 0-19-859520-4 Pbk

Set in 10 on 12pt Press Roman by
Express Litho Service (Oxford).

*Printed in Great Britain by Richard Clay and Company Limited,
The Chaucer Press, Suffolk*

Preface

The primary object of this volume, and those which follow, is to provide a guide to construction technology within the framework of the civil and structural engineering programme of the Technician Education Council. This book covers the first level. In view of current technological developments, however, the scope has been slightly expanded to embrace industrial techniques such as the construction of standardized housing using concrete panels and timber frames clothed in brickwork, now recognized as an effective method of house-building. In order to provide a degree of continuity between chapters, a sketch layout has been inserted as a frontispiece, and this is used as a basis for appropriate examples of construction.

Problems of conversion from imperial measure to SI units still persist. Some materials, e.g. steel reinforcement, are now produced in metric sizes only, but others are manufactured in either metric or imperial dimensions converted into metric, so care has to be taken to avoid error. For this reason, metric equivalents of imperial measurements, including conversion factors, are given at the back of the book. It will also be noted that where long established techniques such as those used in carpentry joints are being superseded by new methods such as metal connectors, emphasis is placed on current practice.

This book, together with those covering higher levels, should prove a useful guide to all technicians of construction whether specializing in architecture, building, civil and structural engineering, or surveying. It is hoped that it will also prove helpful to those now in practice.

R.L.F.

Acknowledgements

Thanks are due to a number of individuals and organizations who have helped in the preparation of this series, particularly the following:

    The Head of the Building and Civil Engineering Department and staff of the Southampton College of Further Education for their helpful assistance.
    Mr J. A. Harvey, BSc., CEng., MICE., FIStruct.E., and staff of Reema Construction Co. Ltd., for their help in industrialized building.
    The National Building Commodity Centre Ltd., and the Building Centre, Southampton, for information and catalogues of which use has been freely made.

Extracts from the following British Standards are reproduced by permission of the British Standards Institution, 101 Pentonville Rd., London, NI 9ND from whom copies of the complete standards may be purchased:

B.S. 459: Part 1, 1954	Panelled and glazed wood doors.
Part 2, 1962	Flush doors.
Part 4, 1965	Matchboarded doors.
B.S. 565: 1972	Glossary of terms relating to timber and woodwork.
B.S. 644: Part 1, 1951	Wood casement windows.
B.S. 990: Part 2, 1972	Steel windows generally for domestic and similar buildings (metric units).
B.S. 1192: 1969	Building drawing practice.
B.S. 1347: Part 3, 1969, AMD 1972	Metric scales.
B.S. 1710: 1971	Identification of pipelines.
B.S. 2028: 1968	Precast concrete blocks.
B.S. 2787: 1956	Glossary of terms for concrete and reinforced concrete.
B.S. 2900: 1970	Recommendations for the co-ordination of dimensions in Building. Glossary of terms.
B.S. 2997: 1958	Aluminium rain-water goods.
B.S. 3589: 1963	Glossary of general building terms.
B.S. 3921: 1974	Clay bricks and blocks.
B.S. 4576: Part 1, 1970	Unplasticised PVC rain-water goods. Half round gutters and circular pipe.
B.S. 4787: Part 1, 1972	Internal and external wood doorsets, door leaves and frames. Dimensional requirements.

Extracts from the following Building Research Establishment Digests are reproduced by permission of the D.O.E. and controller, HMSO:

BRE 128: Part 1   Insulation against external noise.
BRE 140   Double glazing and double windows.
BRE 172   Working drawings.

Meter box installation details in Chapter 15 have been kindly supplied by British Gas and Electricity Boards.

Contents

Abbreviations   xii

A. GENERAL

1. THE BUILT ENVIRONMENT   3
    1.1 Elements of the built environment   3
    1.2 Function and relationship of elements   3
    1.3 Location of elements and the environment   7
    1.4 Considerations affecting choice of site   9

2. CONSTRUCTION ACTIVITIES AND PRODUCTION   11
    2.1 The construction site as a temporary factory   11
    2.2 Drawings and documentation   16

3. DRAWING AND SKETCHING TECHNIQUES   19
    3.1 Free-hand sketches to approximate scale   19
    3.2 Use of scales in free-hand sketching   22
    3.3 Standard symbols and notation   25
    3.4 Production drawings   27
    3.5 Main drawings used in the production process   29
    3.6 Dimensioned drawings   31

4. CONSTITUENT PARTS OF A STRUCTURE   34
    4.1 Substructure, superstructure, and primary elements   34
    4.2 Secondary elements and finishings   35
    4.3 Self-finishes and applied finishes   35

B. SUBSTRUCTURE

5. EXCAVATION WORK ON CONSTRUCTION SITES   41
    5.1 Substructure   41
    5.2 Removal of vegetable soil   41
    5.3 Types of excavation   42
    5.4 Removal of water   46

6. FOUNDATIONS	48
6.1 Function of a foundation	48
6.2 Choice of foundation	48
6.3 Subsoil movement	50
6.4 Primary materials in foundation construction	51
6.5 Pad, slab, and strip foundations	52
6.6 Foundations to beds and pavements	55

C. SUPERSTRUCTURE

7. FUNCTION OF BASIC STRUCTURES	61
7.1 Relationship of superstructure to substructure	61
7.2 Basic concept of a structure	61
7.3 Basic types of structure	64
7.4 Typical structural forms	67
7.5 Component parts of a structure	69
8. FUNCTIONS OF THE EXTERNAL ENVELOPE	73
8.1 The 'external envelope' and its elements	73
8.2 Primary function of the 'external envelope'	73
8.3 Materials for external walls	80
8.4 Solid and cavity wall details	81
8.5 Performance requirements of windows	84
8.6 Conventional types of window	86
8.7 Typical casement-window assemblies	87
9. DOORS	90
9.1 Typical performance requirements	90
9.2 Basic door types	91
9.3 Sizes of doors, doorsets and openings	96
9.4 Typical door frame assemblies	97
9.5 Ironmongery	99
10. BASIC ROOF FORMS	103
10.1 Performance requirements of roofs	103
10.2 Basic roof forms	107

D. INTERNAL CONSTRUCTION

11. ELEMENTAL PARTS OF INTERNAL CONSTRUCTION	113
11.1 Walls, floors, and stairs as primary internal elements	113

12. INTERNAL WALLS: FUNCTION AND BASIC CONSTRUCTION — 118

12.1 Primary function of internal walls — 118
12.2 Loadbearing and non-loadbearing walls — 119
12.3 Typical internal walls and openings — 122
12.4 Finishes to internal walls — 126

13. FLOORS: FUNCTION AND BASIC CONSTRUCTION — 128

13.1 Primary function of ground floors — 128
13.2 Solid ground floors and external walls — 129
13.3 Hardcore — 131
13.4 Suspended timber ground floors and external walls — 132
13.5 Comparison between solid and suspended ground floors — 134
13.6 Suspended timber upper floors — 135
13.7 Strutting suspended timber upper floors — 137
13.8 Floor finishes in domestic construction — 137

14. STAIRS: FUNCTION AND BASIC CONSTRUCTION — 142

14.1 Function of a stair — 142
14.2 Constituent parts of a straight-flight stair — 143
14.3 Critical dimensions in stair construction — 145
14.4 Construction of a straight-flight stair — 146

E. SERVICES AND EXTERNAL WORKS

15. PRINCIPAL SERVICE INSTALLATIONS — 153

15.1 Basic requirements for water, gas, electricity, telephone, and drainage — 153
15.2 Provision in substructure for entry and outlet — 156
15.3 Performance characteristics of basic materials — 160
15.4 Protection of installations — 164

16. DRAINAGE INSTALLATIONS — 168

16.1 Main types of effluent — 168
16.2 Collection and removal of surface-water — 168
16.3 Rain-water eaves-gutters, pipes, and fittings — 170
16.4 Underground drainage systems: general principles — 174
16.5 A typical drainage system — 177

Guide to further reading — 178

Glossary — 179

Index 184

Useful conversion factors 190

Conversion chart at back of book 191

Abbreviations

The abbreviations for units of measurement used in this book conform to those advocated by the International Standards Organization and the British Standards Institution. The imperial equivalents of SI measurements are given in the chart printed at the back of the book in case they are needed. Reference should also be made to other levels; further abbreviations are also given in Figs 3.10 and 16.4.

°C	degrees Celsius	I.E.E.	Institution of Electrical Engineers
b.i.	back inlet	k.d.	knock down
B.S.I.	British Standards Institution	l.b.	loadbearing
B.R.E.	Building Research Establishment	m	metre
		mm	millimetre
B.W.F.	British Woodworking Federation (E.J.M.A.)	m.s.	mild steel
		M	module
C	component	N/mm^2	newtons per sq. mm.
c/c	centre to centre	N.	north
cf.	compare	NTS	not to scale
CI/SfB	A classification system. (*see* Chap. 2)	OG	ogee
		p.e.	plain edge
CP	Current paper	PVC	polyvinyl chloride
C.P.	Code of Practice	r.c.	reinforced concrete
C.S.U.	Consumer supply unit	RHS	rectangular hollow section
c.i.	cast iron	r.w.g.	rain-water gutter
d.p.c.	damp-proof course	r.w.p.	rain-water pipe
d.p.m.	damp-proof membrane	S.	south
dB	decibel	S.A.	satin anodized
dia.	diameter	SS	stainless steel
D.O.E.	Department of the Environment	S.W.A.	Steel Window Association
		s.v.p.	soil and vent pipe
E.	east	TV	television
F.F.L.	finished floor level	t. & g.	tongued and grooved
F.S.	full size	UPVC	unplasticized polyvinyl chloride (rigid)
galv.	galvanized		
GRP	glass fibre reinforced polyester	VC	vitrified clay
		V	volt
G.L.	ground level	VP	vertical plane
HB	hard black	w.c.	water closet
h.r.	half round	W.	west
HP	horizontal plane		

A. General

1 | The built environment

1.1 Elements of the built environment

Any study of the history of building will show that the pattern and design of buildings are influenced mainly by climate, the materials available locally, the basic building method traditionally used, and the way of life of the people of the period. Buildings were originally created to meet primary human needs, mainly the need for shelter, and were constructed by using natural resources such as trees, stone, earth, and water, and by following elementary rules of structural mechanics. The prime purpose has remained unchanged to the present day, i.e. to provide shelter and create a built environment to serve some specific purpose.

With the advance of building science and technology, basic materials shaped from vegetation and minerals in their natural state are now used less frequently, but instead these materials are processed to make products such as hardboard, plywood, concrete, brick, and metal. These seldom resemble their original form, but are transformed to present pleasing or durable exteriors and satisfactory interior spaces for living, storage or the manufacture and processing of materials.

A constituent part of a building is called an *element*. This term also has other meanings, but when related to the built environment and designed for a particular purpose it is called a *functional element*. Walls, floors, and roofs are typical examples. These may be subdivided into components such as internal or external walls, partitions, or joined-up units of floors, walls, or roofs to produce a *structural frame*.

1.2 Function and relationship of elements

A building or structure, by enclosing space, creates an internal environment and the structure itself can be considered as an 'external envelope', the function of which is to provide satisfactory conditions for living and working inside it. The enveloping fabric is also required to fulfil other needs, not the least being to withstand external conditions within stated limits of severity and also to present an image in keeping with its function.

The expectations of modern man are greater than those of his forefathers; so the enveloping fabric must now provide conditions of comfort in terms of warmth, light, ventilation, safety, and security not thought possible even a few decades ago. Modern building regulations insist on standards of weather-resistance, sound- and heat-insulation, fire-resistance, safety, protection, and welfare which often prove irksome to the designer and builder and to the client responsible for payment. The attainment of such standards is not always easy, as these regulations affect not only the materials used but also the building design and construction processes. Even after the building regulations have been complied with, the choice of materials used will further depend on the method of production to be employed, the quality demanded, the accuracy which can be attained in their assembly, and the degree of tolerance permitted in that particular situation. The design of the building and the process of construction will be governed by the nature and sequence of building operations and the relationship between the basic structure and its components. Design and construction will also depend upon the sort of foundation on which the structure will stand and the stresses to which it may be subjected.

The security and protection afforded by the structure and the degree of comfort within its confines at one time depended on massive strength in its fabric, which was achieved mainly by thickness in the walls and the smallness of the windows. Each unit of construction was usually a size that could be manhandled easily and fixed *in situ*. Though slow by modern standards, this method could nevertheless provide a high degree of flexibility.

Contemporary building, in all but its most basic forms, uses elements which have usually been manufactured from products which themselves have been processed from raw materials. In recent years the building industry has become closely involved in the use and development of a wide array of materials, products, and components; such items are normally manufactured by factory processes in a standard range of sizes where possible, and to agreed dimensions and tolerances. Standardization is achieved by setting up machines to perform repetitive functions to predetermined limits of accuracy with tolerances to ensure accurate fit and speedy assembly on site.

DIMENSIONAL CO-ORDINATION. As the mass production of elements and components becomes more common, it becomes very important that they should conform to certain standard dimensions, so as to fit each other. This is particularly true where components have to be incorporated into different assemblies, and because of this the concept of dimensional co-ordination has been evolved. Basically this is the arrangement of the theoretical dimensions of a structure to ensure that these are multiples of a basic unit termed a *module* or M. A modular dimension is often expressed as a multiple of M, e.g. 5M, or it can be a subdivision of it. At one time the brick was used for this dimension but the module has now been standardized at one hundred millimetres (not ten centimetres, as the term 'centimetre' is not used in SI units). The expression *dimensional co-ordination*, when referring to SI units, is the same as *modular co-ordination*. For fuller definitions and methods of controlling dimensions the reader is referred to B.S. 2900: 1970.

Dimensional co-ordination is not confined solely to components; it also relates to the assembly or to the envelope itself. For instance, the designer's size is given as a theoretical dimension usually in whole modules, but the manufacturer must be given some tolerance in his component dimensions. A window, for example, must be made slightly smaller than the opening it will occupy. So a maximum and a minimum limit is given to the manufacturer, usually by stating a mean component size and allowing him a deviation either more or less than the specified mean. This is necessary as no object can be made with absolute accuracy.

There are also other considerations of space requirements, namely those which concern building regulations, floor and ceiling heights, temperature variation affecting warping, joints, and tolerances, as well as submodular dimensions of thin surface finishes or for filling gaps, but these will be considered at a higher level. The basic requirements of dimensional co-ordination considered here are concerned with the manufacture of elements and components within stated limits of deviation (but slightly smaller than the co-ordinating space) and with jointing them in position.

MATERIALS. The composition and character of a raw material normally determines its suitability after processing and often dictates its final shape and size. The properties of steel, for example, render it capable of being rolled or drawn out into continuous lengths, such as in pipes, tubes, sheets, or stanchions; so it is useful in members subject to either tension or compression. As some materials are rigid and others ductile, it is common sense to use factory-made components according to the capabilities of their composite materials. It is essential to keep the stresses in a structure within the safe strengths of the materials of which it is made.

A simple basic product is the *unit* (Fig. 1.1), usually an article made of one material only, such as a sheet of glass, a hardboard panel, or a brick. A *section* (Fig. 1.2) is a term widely used in building and refers to a length of material of uniform section or profile normally shaped by a continuous machine process. Such basic units and sections are often combined to make up *compound units* such as windows and doors in frames of wood, metal, or plastic (Fig. 1.3). Units, sections, and compound units are often built into a structure *in situ* in the traditional manner, but it is also current practice to prefabricate compound units off-site into components made to storey height and width (Fig. 1.4). They are transported to the site by low-loaders and lifted into position by crane, then joined to become part of the structural frame. Such fabricated assemblies are used for both permanent and temporary work.

The interrelationship between elements and components is an important factor in determining the design of a structural frame (Fig. 1.4). An example of this could be a double-decker bus with side-walls, floors, and roof, where each element would not acquire the necessary strength and rigidity until assembled as a framed structure.

6 | *Construction Technology*

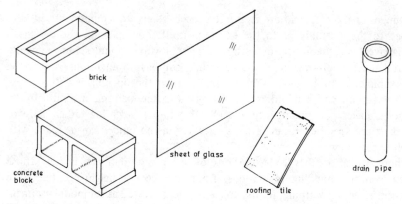

Fig.1.1 Examples of units

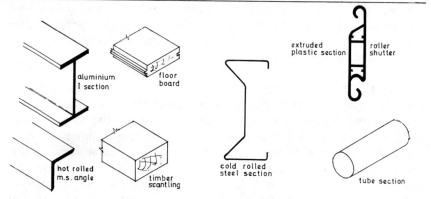

Fig.1.2 Examples of sections

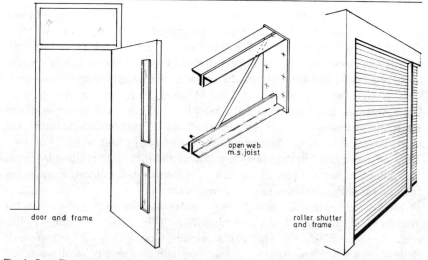

Fig.1.3 Examples of compound units

Size and shape also play a part in design, particularly in the economic aspects of construction. A building square in plan, for example, will enclose a greater area than an oblong one of the same perimeter. As external walls are usually more expensive than internal ones (because of their loadbearing capability, insulation, weather-proof properties, and appearance), the simpler the plan the lower the cost. There are also other economic factors involved, such as the height of the structure, grouping of services to save on capital costs and maintenance, and methods of erection and assembly. The interrelationship between initial cost and subsequent upkeep is a matter which no designer should ignore. These and similar factors will be considered more fully later in this series.

1.3 Location of elements and environment

Although the *in situ* processes known as 'wet construction' are traditionally used, the building of structures also includes the erection and assembly of components brought to site and put together on a suitable foundation. Despite the inventiveness of manufacturers in devising jointing measures and the care taken to adhere to a system of modular and dimensional co-ordination, assembly is still a complex process subject to vicissitudes beyond the designer's control. These often arise because of site location, weather hazards, temperature variation, climatic elements, and other factors, more of which will be discussed in Chapter 8. The purpose of the structure is a prime consideration, together with the type and quality of the materials used and the limitations imposed by cost.

CLIMATE. This has a pronounced effect on the building environment as it is responsible for dispensing sunlight, solar heat, frost, wind, rain, humidity, temperature variation, and other atmospheric conditions known as *weather*. The effect of weather on building elements depends on their location and the severity of exposure, which can sometimes be considerable.

The requirements of a building will depend on the purpose for which it is to be used, which often dictates the kind of materials used in its construction.

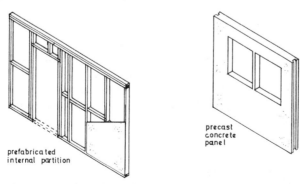

Fig. 1.4   Examples of components

The location of the site however, could play an important part in the choice of the structure. In exposed or windswept areas, a building would be designed to withstand extremes of weather. In a highly industrialized area, on the other hand, a building might have to resist chemical pollution, erosion, noise, and vibration, and the external 'envelope' might have to be made of self-cleaning materials resistant to deterioration and designed to reduce maintenance. Hospitals, teaching institutions, and research establishments would have their own priorities in which environmental considerations could play a considerable part.

SUITABILITY OF COMPONENTS. When using modular grids for fabricated components in which dimensions may be standardized, it is essential that dimensional co-ordination should take account of possible deviations in assembly, even though the elements themselves may be capable of functioning satisfactorily. One of the biggest problems in this respect is movement. Apart from foundations (see Chapter 6) this can arise from such causes as tremor, settlement, subsidence, wind, and change of dimension due to temperature variation. For this reason a great deal of attention is given to the design and insertion of joints. The joint must not only cope with vibration, temperature change, driving rain, thermal expansion, condensation, and sunlight, but must retain its ductility and resistance to heat, damp, and frost over long periods of time. This applies particularly to component assembly joints.

The movement of traditional buildings constructed of unit materials such as brickwork in normal conditions can often be anticipated as the result of experience and does not create the same problems. Expansion joints inserted *in situ* in walls, floors, roofs of traditional materials are not usually subject to quite the same stress and are often constructed differently. The various types of joints will be discussed in later chapters.

PERFORMANCE SPECIFICATION. It has long been the practice of designers and quantity surveyors to specify exactly what shall be put into a structure rather than how the requirements shall be met. It is customary for example, for the designer to require 'an x diameter drainage pipe of b material with c type joints, on a d type bed laid to a y gradient in straight lines from point to point'. By seeking the reasons for these conditions it is possible to arrive at a *performance specification,* i.e. a statement which would define what is needed rather than how the conditions should be met. Such a specification would have many advantages; it would be more rational; it would not become outdated with changing methods and materials, and it would give a guarantee as to suitability of purpose. Elements could be selected which when assembled would have the required performance level, and what is more the junctions could also be made to match this performance. Also the designer could choose elements with the properties necessary to produce the required characteristics.

The implementation of such a performance specification, however, would be a difficult undertaking involving major changes in the entire range of traditional specifications. A new approach to materials testing and method by independent and official testing bodies and government departments would be needed. A start has been made, however, and one or two papers on the subject have been produced by the Building Research Establishment.

1.4 Considerations affecting choice of site

Before a final decision can be taken on the suitability of a site a number of factors have to be taken into account. Some of these result in conflicting requirements, not all of which can be completely satisfied. All considerations will depend on the type of structure required, which may be residential, industrial, administrative, private, or public, and also on the choice of site, which may be governed by economic considerations or merely by personal preference. (The choice could also be due to circumstances beyond the client's control.) The main considerations are:

site location,
economic aspects,
communication,
access,
site conditions,
security.

Site location. This could be influenced by such factors as local authority and planning regulations, convenience, purpose of the structure, type of neighbourhood, etc. Orientation and aspect may have priority or priority could be given to the distance from the main highway, the proximity of amenities, the avoidance of nuisance, or simply personal preference. The client would have to decide which advantages he desired most and what he would be prepared to sacrifice in order to get them.

Economic aspects. Location can also play a role in siting, particularly in relation to facilities available in the area. Good value will be also demanded in terms of space, environment, appearance, and resale price. Initial outlay must be related to running costs, maintenance charges, and efficient operation of the structure and its services, whatever its purpose.

Communication. The choice of site can also be influenced by its need for good transport facilities, which may be important both while the building is being constructed and after it has been finished. The production processes used can be affected by transport conditions including loading and unloading facilities, or the location of a railhead, trunk roads, or terminals with connections to the site. The value of communications will be directly related to user needs.

Access. This is closely related to communication and the function of the structure, but is also concerned with the possible means of approach to the building, taking into account the client's preferences and whether there is right of way. Legal questions could be involved; amenities such as lighting may exist or may have to be provided. Local authorities are usually helpful in such matters.

Site conditions. The designer and client visit the site at an early stage and note such details as could affect the erection and function of the structure. The *soil and ground surface* should be examined and note made of slope and natural features which could affect location or construction. The foundation itself depends on the soil, and this will be discussed in Chapter 5. *Adjacent buildings* could cause obstruction, raise underpinning problems, or affect access; the site surround should be examined for hazards such as vibration from road- or rail-traffic, existing or disused services, underground streams or tunnels, and overhead wires or cables which may have to be diverted or which could be used during or after erection. *Main services* such as water, gas, electricity, telephone, drainage, and waste disposal have to be investigated, permission to connect obtained from local authorities or companies, and adjoining owners may also have to be consulted. This list is not exhaustive.

Security. Note is made of existing boundaries and whether fencing or boarding are needed. Theft and vandalism may have to be guarded against and the possibility of insurance considered. The degree of security needed depends on location. In some cases permanent boundary walls may be built first, especially where these form part of the contract. Internal security is also important.

Finally. It will be seen that functional requirements depend on user needs, which can vary considerably. Where production processes are involved, physical considerations affecting choice of site usually require careful investigation. Questions of local skills, competition for labour, trade union negotiations, facilities for expansion, and similar matters will need scrutiny. Similarly, building for domestic, social, and public use must be considered with regard to amenities which exist or which may have to be provided; these however, are outside the scope of this programme of work. Site activities and layout will be considered at a higher level.

2 | Construction activities and production

2.1 The construction site as a temporary factory

Building is an organizational process which chiefly involves the use of resources such as operatives, materials, machines and similar aids, and the activities of design, manufacture, and construction. The *designer* is mainly concerned with the size and disposition of the space within the 'envelope' of the structure and the services needed to make it operative. The *manufacturer,* if an independent producer of unit components or equipment, is often concerned only with the entire suitability of his product as it leaves the factory, and the *contractor* only with the nature and sequence of erection operations. This could account for the fragmented nature of the industry and the need for closer co-ordination and compromise between those involved. By entering into a package deal with a client, a large contractor sometimes takes full responsibility for both design and execution; this can be successful, particularly in fields where the contractor's reputation is known.

SITE WORK. In its original state the site is often not an ideal place on which to construct a building. It is often unkempt, sometimes derelict, of unsuitable soil, and inconvenient in terms of location, access, necessary services, and building facilities. With the advance of technology, however, building procedure is becoming more rationalized. Much traditional building still remains site-bound and labour-intensive, though the increasing use of components is making it less so. Tradition is still followed in the field of 'wet' construction, however, largely concrete work and brickwork *in situ.* But by incorporating factory-produced units and components into traditional *in situ* masonry and concrete construction, and by employing mechanical plant and equipment, erection time has been speeded up considerably, particularly on work of a repetitive nature.

Sometimes the smaller contractor himself uses part of the site as a workshop or temporary factory for the production of woodwork or pre-cast components, but the industrialization of building methods is now on the increase. This is concerned with both on-site and off-site methods of construction organized in a systematic way in order that erection can proceed as a continuous operation. This is done by the careful planning of activities carried out and by setting up a production line to provide an organized flow of components. The factory area can be independent of the structure itself. By use of multi-functional mechanical plant and by partial replacement of *in situ* work or 'wet' construction with prefabricated units and components, speedy erection can be achieved usually on a much cleaner site. Also the practice of producing standardized 'off the peg' components in a factory and delivering them ready for fixing is now commonplace.

Whether traditional or industrialized, on-site organization of materials, components, and labour is vital in construction procedure; pre-planning of each stage is essential, and adequate time must be allowed for working out details before operations can commence. Pre-planning of activities should cover:

layout of the site,
the sequence of work,
design, manufacture, and fixing of standardized components,
mechanical plant.

Layout of site. This must permit an orderly flow of activities, with good access, enough working space, and direct routeing and turning areas for each of the operations. Stacking and storage facilities with adequate access must be allowed for and avoidance of double handling is essential. Covered areas for workshops and welfare amenities may have to be provided, as well as hard-standings for vehicles. On the basis of these requirements a block plan should be drawn up to show the position of the structure and layout of all temporary buildings, services, outlets, and areas. Building regulations, Factories Acts, and by-laws may have to be consulted and requirements of local authorities met.

Should a layout for repetitive component production be required, it is wise for a visit to be arranged to a well-equipped concrete or similar factory which specializes in proprietary products or components for site installation. The general arrangement of offices, stores, testing laboratories, casting bays, workshops, and curing and storage areas should give a good indication of what is required.

Sequence. Whatever type of structure is envisaged, continuity of production is essential. A programme of work must be drawn up to ensure an orderly sequence of operations with repetition of activities where possible in order to create a routine effect; this will help to preserve sequence, save time, and increase output. Working in series by means of repeated performance is the most efficient way to accomplish this and the work must be broken down into stages to enable this to be done.

Standardization. Where skilled labour is in short supply and the structure is designed to permit the repetitive manufacture of its components, prefabrication can prove economical. Apart from production methods, however, other factors have to be considered; fixing and jointing techniques, for example, are of the utmost importance. They must be as simple as possible with due allowance for tolerance and variation, yet remain quite watertight. Lifting methods must be worked out, with loops, dogs, or shackles built into the components at points of equilibrium to simplify placing. Some of these may have to be removed by cutting or burning after fixing so they must be concealed. These are normally of m.s. rod but permanent fastenings are rust-proof, usually stainless steel.

Mechanical plant. To assist in the speedy operation of site work and reduce the amount of manual work needed, a wide range of mechanical plant is available, from small hand-tools to large mobile equipment. Powered hand tools are now used in great variety including poker vibrators, tampers, saws, drills, hammers, sprays, pumps, etc. Larger plant includes power barrows, drag shovels, scaffold cranes, calf-dozers, compressors, concrete-mixing plant, and much besides.

DESIGN RELATED TO CONSTRUCTION PROCESS. Units and components may be manufactured either in factories which will produce constituent parts ready for assembly on site or, as explained above, the site itself may be adapted as a temporary factory. In either case each step in the production is worked out on the drawing-board together with the specification, materials testing, structural calculations, and fixing method. Factory-made products, by and large, are of consistent quality and can be competitive provided the output required justifies the capital outlay. A site factory, though often less well-equipped and more prone to interruption of production, can nevertheless prove cheaper and at the same time possess the advantage of having both manufacture and fixing of components under the control of one person.

Frequently both systems are used simultaneously, though this needs to be carefully planned at the design stage. Much construction work at one time performed on site is now factory-made as elements or components, and delivered to site for assembly and fixing. These include preformed panels of pre-cast concrete or sometimes brick, or various forms of cladding or curtain walling complete with built-in door- and window-frames. Formwork for *in situ* site concrete may also be made up into panels or shutters and can be of timber, steel, or combinations of these. Site casting of elements on larger sites is also done especially where lifting-plant is available; walls, floors, and roof panels can be pre-cast and fixed in this way. Prefabrication of constituent parts of a

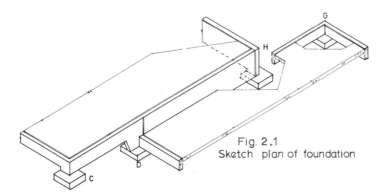

Fig. 2.1
Sketch plan of foundation

Location

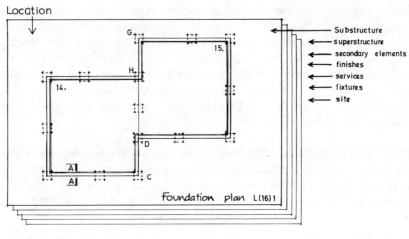

Assembly

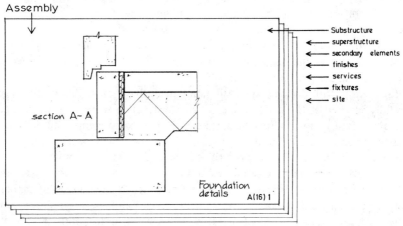

Components

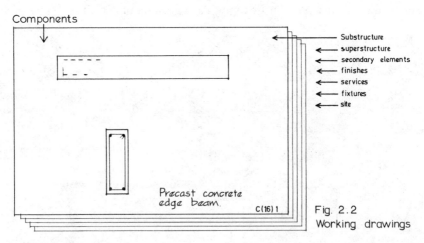

Fig. 2.2 Working drawings

Construction activities and production | 15

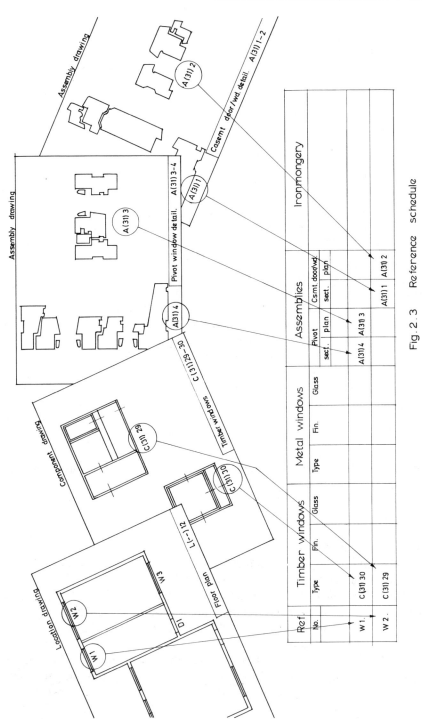

Fig. 2.3 Reference schedule

component is also common, as in the use of reinforcement cages made up and positioned before concrete is poured.

Once the structure is sufficiently advanced, e.g. when the skeletal frame is in position, it sometimes pays to shroud the carcass in translucent plastic sheeting thus turning the enclosed space into a temporary factory. On special contracts where time is valuable this can save production time. And in any case it provides storage facilities, security, comfort, and protection – particularly useful during the winter months.

2.2 Drawings and documentation

Drawings conveying technical information concerning a project need to be systematically classified in order to make location easy and to enable information to be retrieved without difficulty. This problem has been given much thought by the B.R.E., which, after some research, reported its findings in a Current Paper C.P. 18/73 on which the following account is largely based. The system they recommend is now being used by a number of firms and organizations, often in a form modified to suit individual needs and priorities.

In considering such a system it is necessary to keep in mind what the builder needs to know. He clearly requires information on the shape, size, and location of the building and all its constituent parts; he also wants details of the composition of the materials and jointing or fixing methods. Information on the shape, size, and location is normally given on working drawings, and reference to these may be made by means of grids on plans and sectional and elevational keys. Such references must be repeated on details to pinpoint location in the building. Materials to be used are usually stated on the drawings or relevant schedules; references are also made by means of graphic symbols (see Figs 3.9 and 3.10).

In order to classify information for easy retrieval it is usual to separate drawings into the broad divisions of *location, assembly,* and *components.*

Location drawings consist of block plans, site plans, or similar layouts which show the position of buildings in relation to known reference-points and which indicate general construction and location of principal elements, components, and assembly details.

Assembly drawings shows the construction of buildings and details of junctions between elements, or components, or both.

Component drawings show basic sizes, systems of reference, and all information necessary for the manufacture of items and their application.

Schedules provide a good method of gathering together repetitive information and these are usually stored as a separate group (Fig. 2.3). The object of subdividing drawings into these categories is to rationalize filing and make retrieval simpler. Other groupings are also in use. The list of elements selected must represent a well-defined division of the project and should avoid ambiguity and doubt. Some organizations prefer to use their own system, but many follow the universal system of elemental breakdown known as CI/SfB classification.

UNIVERSAL CLASSIFICATION. The CI/SfB classification for basic design and performance criteria is firmly established for components and materials and is now used almost exclusively. The subject of CI/SfB will be discussed at another level but an elemental breakdown is given in Table 2.1. When numbering drawings it is customary to prefix the CI/SfB code reference with the letter L (location), or A (assembly), or C (component). A simple numbering system is thus obtained giving essential title information which may also be used for cross-referencing (Fig. 2.2). When using the system it is necessary to title the complete set of drawings for the project in accordance with the code. Titles should be brief and just sufficient to identify the subject.

NUMBERING OF DRAWINGS. To understand how the system works it is necessary to study the block plan (Fig. 3.13) and also the frontispiece which shows houses numbered 14 and 15 (enlarged in Fig. 2.2). A sketch of the substructure is also given in Fig. 2.1. The working drawings of the house foundation are shown in Fig. 2.2 with the location plan shown first followed by the assembly drawing, and lastly by the component drawing. There would, of course, be additional location drawings to deal with the project as a whole. The assembly drawing here shows the foundation detail, and the component drawing gives the pre-cast concrete edge-beam detail. Similar location, assembly, and component drawings would be prepared for the superstructure, services, internal construction, fittings, and external site works. Each drawing would be numbered in accordance with the CI/SfB breakdown (Table 2.1). Should two or more elements be shown on the same drawing then the highest number would be chosen; thus the elemental CI/SfB number for this foundation would be L(16) *foundation* and not L(13) *floorbeds*. The final figure in the code L(16)1 refers of course to the sheet number.

Cross-referencing. Searches for information usually proceed from the general to the particular, and this is done by means of schedules as shown in Fig. 2.3. The schedule opens with reference to the location drawing which contains details of the relevant elements, components, or parts. The appropriate component drawing numbers are then added to the schedule — in this case C(31)29 and 30 — followed by the assembly number A(31)1 to 4. It will thus be seen that drawings relating to (in this case) the position of the window opening, type of window, and method of assembly can all be traced by reference to the schedule. It is necessary, however, that all users, including designers, manufacturers, and contractors, should adopt the same numbering system, and to avoid misunderstanding a written explanation should accompany each set.

Table 2.1 Elemental breakdown (from CI/SfB Table 1)

(– –) Site, project

Substructure	Superstructure				Services		Fittings		Site
	(2–) Primary elements	(3–) Secondary elements	(4–) Finishes		(5–) Mainly piped	(6–) Mainly electrical	(7–) Fixed	(8–) Loose	(9–) External elements
(1–) Ground, substructure									
(10)	(20)	(30)	(40)		(50)	(60)	(70)	(80)	(90) External works
(11) Ground	(21) External walls	(31) External openings	(41) External		(51)	(61) Electrical supply	(71) Circulation	(81) Circulation	(91)
(12)	(22) Internal walls	(32) Internal openings	(42) Internal		(52) Drainage, waste	(62) Power	(72) Rest, work	(82) Rest, work	(92)
(13) Floorbeds	(23) Floors	(33) Floor openings	(43) Floor		(53) Liquid supply	(63) Lighting	(73) Culinary	(83) Culinary	(93)
(14)	(24) Stairs, ramps	(34) Balustrades	(44) Stair		(54) Gases supply	(64) Communications	(74) Sanitary	(84) Sanitary	(94)
(15)	(25)	(35) Suspended ceilings	(45) Ceiling		(55) Space cooling	(65)	(75) Cleaning	(85) Cleaning	(95)
(16) Foundations	(26)	(36)	(46)		(56) Space heating	(66) Transport	(76) Storage, screening	(86) Storage, screening	(96)
(17) Piles	(27) Roofs	(37) Roof openings	(47) Roof		(57) Ventilation	(67)	(77) Special activity	(87) Special activity	(97)
(18)	(28) Frames	(38)	(48)		(58)	(68) Security, control	(78)	(88)	(98)

18 | *Construction Technology*

3 | Drawing and sketching techniques

3.1 Free-hand sketches to approximate scale

Much technical information, apart from the written word, is imparted by means of drawings and to interpret them it is necessary to be able to identify the main types used, to recognize standard graphic symbols and notation, and to prepare production drawings using instruments and metric scales (Fig. 3.14). It is also essential to be able to communicate graphically by means of free-hand sketches drawn to approximate scale.

FREE-HAND SKETCHING. Before free-hand sketching can be attempted it is fundamental that some ability in drawing lines, both straight and curved, should be acquired. They should be firm, clean, and of even thickness. (A scale should never be used for drawing lines.) An HB pencil is probably the best to start with, preferably a clutch pencil which can be sharpened by using the removable clutch knob at the top. When guide-lines, either horizontal or vertical, are being drawn on a sketching pad or note paper, they should be made parallel to the edge of the paper, and kept straight by using the little finger as a guide where possible. A free-hand line should be drawn in one movement from end to end. Curves are best drawn upward in a smooth continuous line by holding the pencil a little way back from the point. Free-hand circles are usually drawn in two halves using guide lines with radius points marked on.

Some practice is then necessary in drawing geometrical figures such as squares, rectangles, triangles, circles, or combinations thereof. Squares and rectangles should be checked by the length of their diagonals and all figures drawn in proportion. Continuous and connecting curves such as ogees should then be drawn including straight lines at intersections such as tangents (Fig. 3.1a). Polygons should then be tried, especially hexagons drawn symmetrically as an aid to isometric drawing (Fig. 3.1b). As we can see in this figure, to draw an angle of 30° free-hand is fairly easy: cb is equal to half ac and this distance can be judged by eye. Hexagons should be considered as six equilateral triangles, and some skill in drawing these and other polygons both inside and outside circles and vice versa, should prove very useful.

ORTHOGRAPHIC PROJECTION. This is a method of showing solid or three-dimensional objects on paper by means of plans, elevations, and sections. In Fig. 3.2, an object is shown suspended in space. When looking down, the plan is seen below the object on the horizontal plane; looking at the front, the

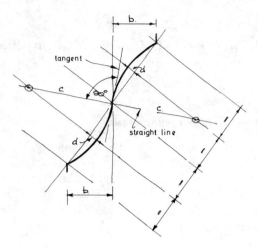

Curves and tangents Fig. 3.1a.

elevation is seen behind the object on the vertical plane; a side-view from the left is also shown. By opening out the sheet, the plan, elevation, and side-view may be seen in one plane. This arrangement is known as *orthographic projection*. The object may be sliced through either horizontally or vertically to show sectional views. Side-views are not usual, sections being preferred. Sections may be shown from either side, provided the pointers on plan or elevation show the direction of view. A standard method of showing planes of section is given in Fig. 3.11, but draughtsmen frequently use their own.

ISOMETRIC PROJECTION. This is used extensively to produce a three-dimensional effect as shown in Fig. 3.3. The horizontal lines shown in orthographic projection are drawn (usually) at 30° in isometric drawings. Dimensions may be scaled or measured by paper strip direct from the orthographic drawing, if one exists,

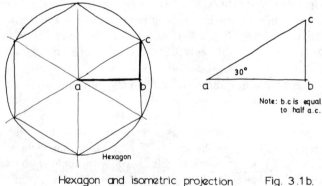

Hexagon and isometric projection   Fig. 3.1b.

Drawing and sketching techniques | 21

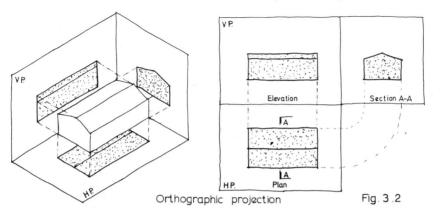

Orthographic projection   Fig. 3.2

to get the right proportions, though only horizontal and vertical measurements can be transferred in this way. The method will be made clear from Fig. 3.3 which shows a cube in both isometric and orthographic projection. It should be noted that the outline of the isometric cube is a hexagon. A further example of isometric and orthographic projection is given in Fig. 3.4. For simple sketches the orthographic drawing can be dispensed with.

Circles in isometric. The simplest way for a beginner to draw a circle in isometric without instruments is shown in Fig. 3.5. First a square is drawn freehand and a circle inscribed. The square is then divided into (say) six divisions. Next, a cube is drawn in isometric using the square side AB. The vertical guidelines are then transferred by using the distances and points a–b and b–c as shown. When sufficient points have been transferred, the circle can be drawn in.

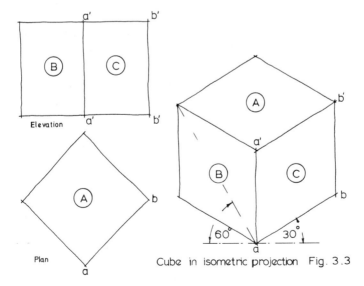

Cube in isometric projection   Fig. 3.3

22 | Construction Technology

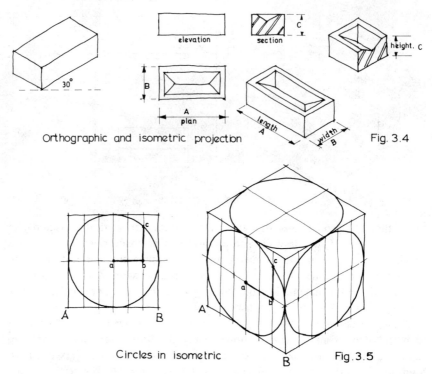

Orthographic and isometric projection

Fig. 3.4

Circles in isometric

Fig. 3.5

The isometric circle is in fact an ellipse, and the major axis has a distinct relationship to the minor axis. After a little practice most of the guidelines can be dispensed with. A necessary point to remember is that the curves always touch the straight edges tangentially.

3.2 Use of scales in free-hand sketching

Scales can be very useful when free-hand sketches are required (Fig. 3.12). They can be used to transfer dimensions directly from drawings and to decrease or increase the size of the sketch by changing scales. They can also be used as invisible guidelines for keeping the sketch vertical or at the correct angle, without themselves being drawn in.

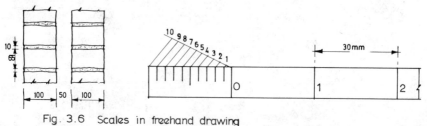

Fig. 3.6 Scales in freehand drawing

Standard metric scales, however, are not always suitable. The size of a sketch sometimes depends on the size of the sheet available and on the amount of detail required to be shown. After some practice in scale-drawing it will be possible to visualize a unit dimension of 100 mm or 300 mm, and, from these, the size of the finished sketch. A 100 mm unit is popular for drawing as it is about the size of half a brick; this can be used to keep other dimensions in proportion (Fig. 3.6).

It is quite easy to make one's own scale and this is often used to enlarge or reduce an original drawing or object. To enlarge an original three times, for example, first divide a line into five or six divisions each 30 mm long; then subdivide the first division into ten parts as shown. By using a standard 1:1 scale (Fig. 3.12), to measure the actual size of the original, the enlargement can be drawn using the new scale. The reverse process can be used to reduce a drawing. The easiest way to divide a line into any number of parts is to use a standard scale held at point 0 at any convenient angle, and mark off the divisions — ten in this case. These are then projected on to the base-line by means of parallel lines (Fig. 3.6).

To divide a space into any number of equal parts all that is necessary is to place a standard scale across a gap and mark off the number of divisions required (Fig. 3.7). An application of this is shown in the drawing of a tusk tenon joint as described in Chapter 13.

Opening in a garden wall. Adopting the method just described, free-hand drawing in proportion is not difficult and orthographic projection can be used to any convenient scale once the construction is understood. An example of this is shown in Fig. 3.8. First, the necessary detail is decided upon, together with the size of the finished sketch. The procedure is then as follows.

Assume the total height of the actual opening to be 1800 mm and width 900 mm. Draw a centre line free-hand and allow enough room for a plan and section. Draw a rectangle in elevation, two parts high and one part wide, to the required size and check the diagonals as shown. Divide the width into four

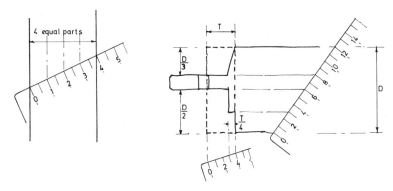

Fig. 3.7   Division of spaces

parts on plan and each part will represent 225 mm or one brick unit. Mark these brick lengths on to the edge of a sheet of paper to give the scale; then divide one brick into two to give half bricks and another into three to give brick heights, i.e. four-bricks to 300 mm.

To draw a semicircular arch, mark off the radius plus half a brick on a paper edge and use this to draw the elevation; then mark off brick heights around the arch intrados and radiate lines to illustrate voussoirs. Divide the height from the ground to the arch springing into brick heights. Do this a few times vertically across the elevation in order to keep the courses horizontal. To keep both halves of the arch similar, a good idea is to use tracing paper. By tracing one half and turning the paper over, the outline impression can be used to draw the other side.

The isometric sketch can be drawn by using the same paper scale to transfer horizontal and vertical distances from the orthographic sketch. The arch should present no problem provided that both intrados and extrados are drawn as described for Fig. 3.5 and that both are ellipses. (A line parallel to an ellipse is not an ellipse.) As can be seen, such sketches are excellent for showing detail

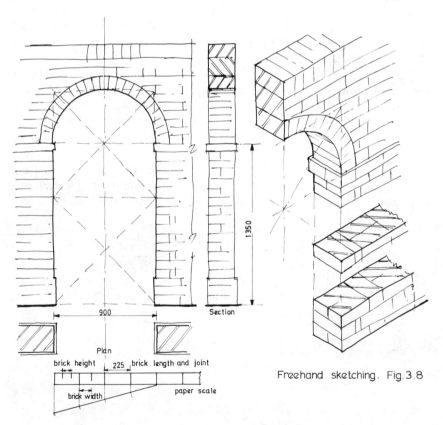

Freehand sketching. Fig. 3.8

Drawing and sketching techniques | 25

on plan or section. Apart from brick sizes, other basic units can be used, such as concrete blocks, door openings, room sizes, storey heights, etc. Open-air sketching of elevations involves other techniques and will not be described here.

3.3 Standard symbols and notation

In order to indicate materials and details on production drawings, conventional methods of illustration are used as recommended in B.S. 1192. The main types are shown in Fig. 3.9 and are usually represented on plan and section only. An exception is the screed symbol which may also be used to show stucco in elevation. Drawings shown in section usually have a thicker outline, and the conventions shown here will be used throughout the book together with others as the need arises.

GRAPHIC SYMBOLS. These are widely used in construction drawings, and a sample of those in general use is shown in Fig. 3.10. There are others, and the designer often has to prepare his own for specific purposes, in which case he

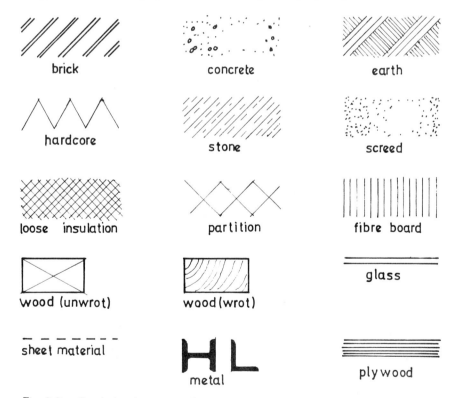

Fig. 3.9 Symbols for materials

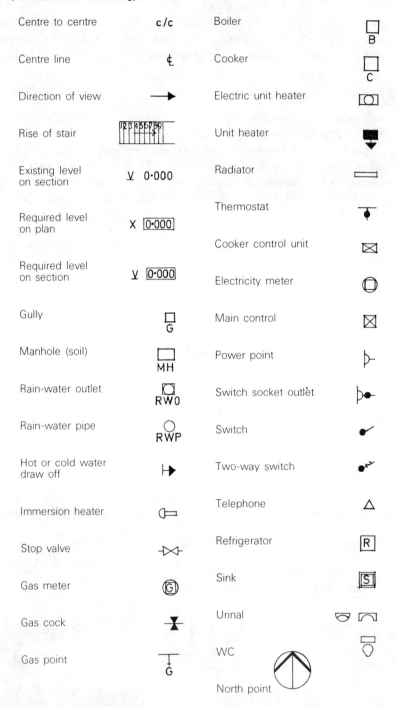

Fig. 3.10 Graphic symbols

will describe them in the notes on the layout sheet (Fig. 3.11). A full list of conventional symbols is given in B.S. 1192, Building Drawing Practice.

NOTATION. Reference has been made here only briefly to dimensional and modular co-ordination and the CI/SfB system, but there are many excellent publications dealing with standard notation and metrication. The standard notation for lines and dimensions is shown in Fig. 3.11.

Decimal points used in drawings and printed and typewritten work are normally placed on the line, e.g. 6.5. When the value is less than unity the decimal point should always be preceded by zero, e.g. 0.65. B.S. 1192 also recommends that dimensions on drawings should be written as follows:

whole numbers indicate millimetres, e.g. 6500; and

decimal expressions to three places indicate metres, e.g. 6.500.

The conventional way of showing linear dimensions on drawings is:

      6.5 mm
     65
    650
   6500
  65 000

Where notes, explanations, and descriptions are used, it is customary to add 'mm' or 'm' after a measurement to make its meaning clear, e.g. 120 mm or 6.500 m. Where five digits are used, as above, a thousands marker is shown by means of a thin space as shown.

Sequence of dimensions on drawings, schedules, and similar documents should always be consistent and expressed usually as *length* followed by *width* and then *height*. Expressions such as 'breadth', 'depth' should be confined to simple components in situations where confusion is not likely to occur. 'Thickness' is normally reserved for sheet materials.

3.4 Production drawings

Production drawings, or 'working' drawings as they were once called, fulfil the need to communicate both adequately and accurately the information necessary to enable the work to be carried out. An important consideration is that of scale, which will be influenced by a number of factors. There is first the need to achieve economy of effort and time in preparation of drawings; then the character and size of the subject must be considered and also the desirability of keeping the sheets for one project to one size as far as possible. Large drawings may be folded where necessary as recommended in B.S. 1192.

With the advent of modular and dimensional co-ordination, the use of printed grid paper is increasing and a number of grid sizes is now available. These are based on a 2 m grid size which satisfies most of the requirements of the different categories of drawings and the preferred scales listed below.

28 | *Construction Technology*

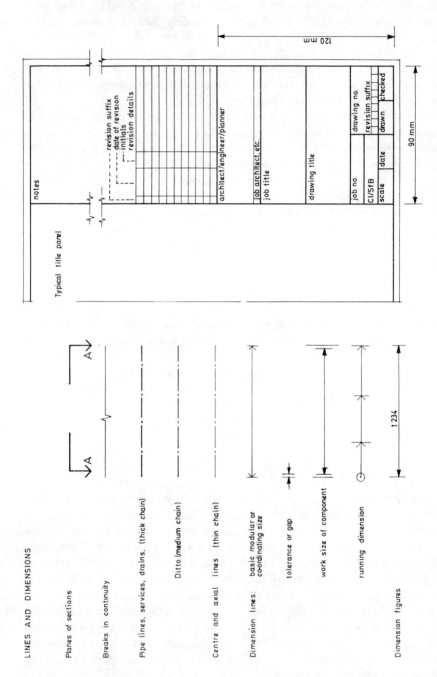

Fig. 3.11 Standard layout and notation

PREFERRED SCALES. The standard types of metric scale prepared according to B.S. 1347, are now in general use and are shown in Fig. 3.12, with a comparison line giving the relationship between the four grades. The scales 1:1250 and 1:2500 are included to satisfy the requirements of the Ordnance Survey who have certain problems in the change-over to preferred scales. The standard scales for production stage drawings are listed in B.S. 1192 and are as follows:

Location drawings

Block plan	1:2500	(0.4 mm to 1 m)
Site plan	1:1250	(0.8 mm to 1 m)
Site plan	1:500	(2 mm to 1 m)
	1:200	(5 mm to 1 m)
Gen. location	1:200	(5 mm to 1 m)
	1:100	(10 mm to 1 m)
	1:50	(20 mm to 1 m)

Component drawings

Ranges	1:100	(10 mm to 1 m)
	1:50	(20 mm to 1 m)
	1:20	(50 mm to 1 m)
Details	1:10	(100 mm to 1 m)
	1:5	(200 mm to 1 m)
	1:1	(full size)
Assembly	1:20	(50 mm to 1 m)
	1:10	(100 mm to 1 m)
	1:5	(200 mm to 1 m)

3.5 Main drawings used in the production process

The main types of drawings used in connection with building are those described under preferred scales. Surveys and layouts also include maps, town surveys, etc., which are not much used in the construction process apart from block and site plans (Fig. 3.13).

Landscape drawings, though concerned with the building environment and external works, are not closely connected, apart from overlapping areas such as retaining walls, footpaths, and small bridges.

Measured drawings are usually the province of the architect, normally prepared from existing buildings. These are often of educational or historic interest but can also be required for modernizing outdated structures or for retrieval purposes.

PRODUCTION DRAWINGS. These will be the main concern of the technician. They are usually prepared on tracing paper or film almost always in ink for reproduction as a contact print to the same scale. They may be location or key

30 | *Construction Technology*

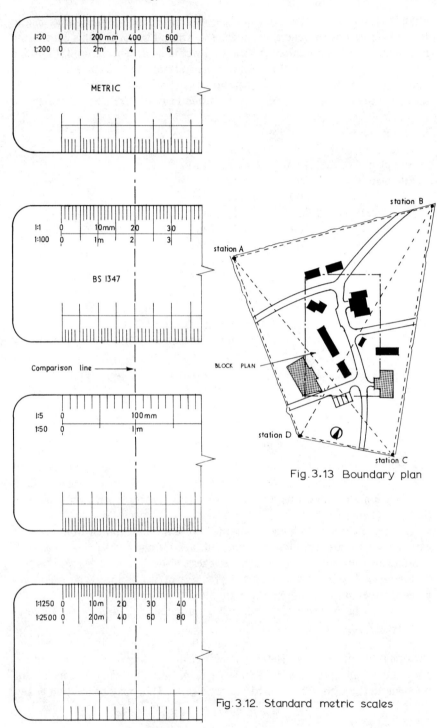

Fig. 3.13 Boundary plan

Fig. 3.12. Standard metric scales

Drawing and sketching techniques | 31

drawings, assembly drawings, component ranges, or component detail drawings. The latter are often supplied by the manufacturer sometimes in catalogue form and usually contain detailed fixing instructions. The standard types of drawings are:

Block drawings	Site outline of new buildings, existing buildings, reference grids, dimension lines, etc. (Fig. 3.13).
Site drawings	Site outline as above but with general detail added.
General location drawings	Primary functional-elements such as horizontal and vertical sections of loadbearing walls, structural slabs, etc. (Fig. 2.3).
Component drawings	Secondary elements and components in horizontal and vertical sections such as non-loadbearing partitions, windows, doors, and also components in elevation (Fig. 2.3).
Assembly drawings	General location drawings as above but which contain all information relating to assembly such as full size details, junctions of primary elements and methods of fixing. They also include manufacturers' fixing instructions (Fig. 2.3).

3.6 Dimensioned drawings

These are broadly of two kinds, *traditional* and *grid*. Traditional drawings usually show plan, elevation, and section set out in conventional form based on orthographic projection and dimensioned as shown in Fig. 3.11. There are many variations however. Fig. 3.14 shows how dimensions can be inserted clearly on one sheet for the convenience of the technician or site agent. This represents a standard base for an end-of-terrace type of prefabricated house with the position of services located in the slab.

Such a plan could also have been shown as a location drawing on a grid. This method is popular with manufacturers of industrial systems where components are of standard size and also for structures where layouts conform to modular co-ordination. Grid notations, when applied to planning and structural grids, usually use a letter/number system of reference which defines grid lines on one axis by letters and the other by numbers. This method will be considered further at another level.

A typical method of dealing with details is to draw assemblies full size, as could be done with those shown in Fig. 2.3 Only overall sizes need be given as the rest can then be scaled. It must be remembered, however, that when dealing with timber frames, allowance has to be made for the difference between the modular basic size (before planing) and the work size (actual dimension after

Construction Technology

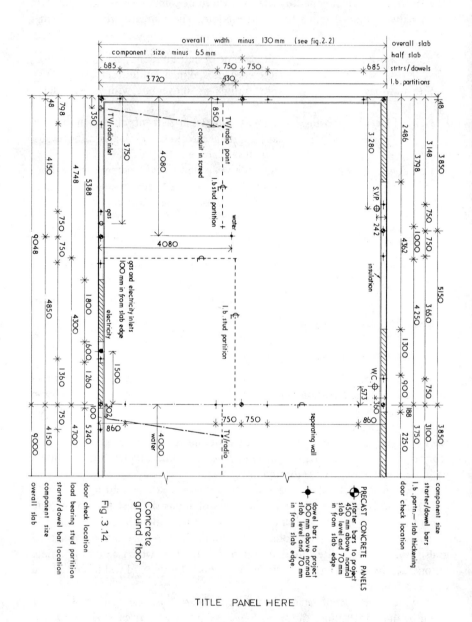

Fig. 3.14. Concrete ground floor

planing). Once the component has been made it has then to be assembled into the structure. Typical assembly details are given in Fig. 2.3.

TITLE PANELS. The layout of drawing sheets must be done in a systematic manner and is usually based on an established pattern designed to suit the particular organization or firm. Certain basic information is necessary however. This includes job titles; subject of drawing; scale; date of drawing; job number; drawing number; revision suffix and SfB reference as mentioned in Chapter 1. A typical panel based on that shown in B.S. 1192 is given in Fig. 3.11.

Orientation of plan. Every location should show a north point, which the designer usually devises for himself. The standard design shown in B.S. 1192, reproduced in Fig. 3.10, is used though some draughtsmen prefer to add an 'N' to avoid risk of error.

4 | Constituent parts of a structure

4.1 Substructure, superstructure, and primary elements

A *structure,* as referred to in Chapter 1, consisting of an organized combination of connected units, can be constructed to perform a function requiring some measure of rigidity, but it need not necessarily provide shelter from the weather; that is the function of a *building.* But with the introduction of Portland cement concrete, a building or structure is not now usually thought of as being composed of units as it was centuries ago when the early cathedrals were being built, or when walls and columns supported lintels of stone. Today, buildings or structures are considered as parts or *elements* of construction each having its own functional identity such as a foundation, floor, wall, or roof.

SUBSTRUCTURE. The structure itself is normally regarded as consisting of a *substructure* and a *superstructure.* The former is identified as that part of a building or structure below the level of the adjoining ground. In traditional building it has long been the practice to consider the substructure as reaching up to the damp-proof course, which is conventionally placed at 150 mm above ground level. A substructure refers to all parts of the understructure including the *foundation* whose function it is to spread loads applied to the supporting soil (Fig. 2.1). The foundation pads shown in that drawing are part of the substructure when taken as a whole. Substructures may also be seen in Figs 7.1a and 7.1b as those parts extending up to ground-floor level. The ground floor itself, however, is often considered to be part of the superstructure especially if suspended (see chapter 13).

SUPERSTRUCTURE. This item normally applies to all structural work above the substructure and comprises such elements as external and internal walls, both loadbearing and non-loadbearing, the structural frame, ground- and upper-floors, stairs, and roof. Referring to Figs 7.1a and 7.1b, the superstructure is that part of a building above the ground. Further examples of superstructures can be seen in Figs 7.5e to 7.5j.

PRIMARY ELEMENTS. These are so called because of their functional requirements, and usually relate to the main walls, floors, and roof. In addition to performing such roles as enclosing or dividing space, they are also called upon to provide adequate strength and support, and to meet such demands as resistance to weather, fire, and noise; also to meet safety and other requirements in industrial buildings, or to maintain whatever degree of comfort in dwellings is required.

4.2 Secondary elements and finishings

Secondary elements normally refer to internal walls and partitions, particularly when wholly non-loadbearing (Chapter 12), and also to stairs (Chapter 14), but the distinction between primary and secondary elements, and secondary elements and finishings is often blurred. Partitions, for example, can vary from lightweight hollow block of sound construction (Fig. 12.3), to framed plate glass, which could also be considered a finishing.

FINISHINGS. It has been pointed out that the difference between a structure and a building is that the structure can be complete in itself when it is only a carcass, i.e. a building which is structurally complete but otherwise unfinished; though a finished building can also be called a structure. To convert a carcass into a complete building (excluding services) it is necessary to apply a casing. This can be a preformed cladding, or traditional construction with appropriate treatment of surfaces as described below. Fixtures and fittings are considered as finishings when treated as functional, as are decorative objects such as joinery work when fixed to the carcass rather than built into it.

4.3 Self-finishes and applied finishes

Some designers like to make a distinction between *finishings* and *finishes* by considering the former as a casing and the latter as treatment of the casing. Curtain walling or cladding as an external facing could be considered as an example of the former. It is difficult to be definitive about this as there is bound to be some overlapping.

Building or structural finishes are normally effected by self-finish or by application. The most common example of self-finish is to be seen on concrete in any of its many forms such as *in situ* concrete, artificial or reconstructed stone, pre-cast concrete, etc. Although the term could be applied to the many preformed products, such as processed timber, plasters, plastics, etc., it is normally reserved for *in situ* processes such as pouring concrete into preformed moulds or specially lined formwork to produce a desired result. Two examples of the latter are given in Figs. 4.1(a) and 4.1(b).

INTEGRATED FINISHES ON CONCRETE. Mould liners used for obtaining special finishes can be made from many materials. Timber, steel, glass fibre, rubber, hardboard, foamed polystyrene and other types of plastics are in general use. The design is applied in reverse to the face of the liner or formwork by screwing, nailing, or glueing. Sometimes the formwork itself is of moulded plastic; one type is in wide use for constructing reinforced-concrete floors. Almost any

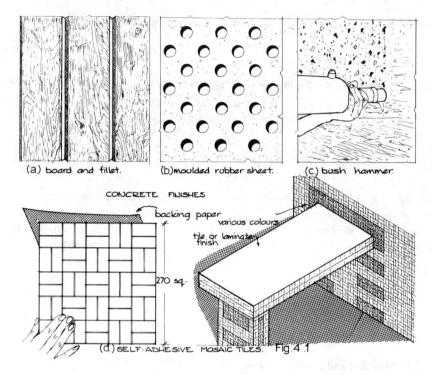

CONCRETE FINISHES
(a) board and fillet. (b) moulded rubber sheet. (c) bush hammer.
(d) SELF-ADHESIVE MOSAIC TILES. Fig. 4.1

shape, texture, or colour can be had from a combination of the mould shape and the materials used in the concrete.

Once the formwork is struck, further treatment can be applied by means of powered hand-tools to provide a rough or smooth texture, or different effects can be obtained by blasting, spraying, scraping, brushing, or by application of etching acid. Fig. 4.1(c) shows one such finish. This treatment is popular as it tends to remove or disguise any blemishes on the face of the concrete, such as board marks, construction joints, irregularities, and uneven texture.

PREFORMED FINISHES. These are often created in the product as it is being manufactured and applies to a whole range of units and components as well as thin surface finishes. Pre-cast concrete work is an obvious example. Not only can the shape be created but the surface texture may be applied to the mould as a rendering before the concrete is poured. The surface may then be treated if necessary after the casting is removed from the mould. This method is widely practised on materials other than concrete, particularly when acids are applied for an etched finish. Plastic coatings in the form of a paste solution sprayed on as a film now provide a wide selection of finishes supplied ready for fixing. Preformed finishes are also used as external facings to cladding but panels and infilling will be dealt with at another level.

APPLIED FINISHES. The main thin surface finish in this group is painting. However, any form of thin surface material that is secured by mechanical or adhesive means comes into this category (Fig. 4.1(d)). Applied finishes are often in the form of preformed coverings and include wall-, ceiling-, and floor-linings of paper, metal, wood veneers, plastics, panelling, casing, wall tiling, and the rendering, plastering, and pointing of brickwork, etc. The term can cover cladding of the structure, roofing materials, either functional or decorative, and in fact any covering material of a permanent nature.

FINALLY. It will be appreciated from the foregoing that the subject of components and finishes is a wide one and is involved in some way with every element and compound both functional and decorative. Only the main categories of finishes have been mentioned here but further examples will be given as appropriate in the following chapters.

B. Substructure

5 | Excavation work on construction sites

5.1 Substructure

As described in Chapter 4, a substructure is that part of a building or structure which is below the level of the adjoining ground and of which a foundation is a part. Its design will be affected by the kind of soil on the site and the loads it will safely carry.

Before commencing site work it is necessary to know the type of structure which is to be built and the work that has to be carried out. The structure may require a shallow or deep strip foundation, base pad, short-bored piling, basement work, retaining walls, drainage work, stepped foundation, or work on a sloping site, or a combination of any of these. Substructure foundations for framed structures may involve column bases or they could be in the form of a floating concrete raft. The substructure must be considered in the light of both the nature of the ground and the load it will have to carry.

The location of the site, its nature and position relating to adjoining properties should also be studied before work is begun. Some buildings will need special treatment such as shoring up, underpinning, or strengthening of existing foundations. However, only the most straightforward of excavation problems will be dealt with here. A site plan showing the location of all known services including drains, manholes, wells, old workings, basements, etc., should be in the site agent's hands before starting. Some contractors engage surveying firms to carry out this work.

5.2 Removal of vegetable soil

Preparatory work on site clearing will depend of course on the nature of the ground. The site may be of virgin soil, grass, farm land, woods, orchards, scrub, or brush. Wet conditions can sometimes be detected by the type of vegetation, and evidence of landslip shown by broken ground around hills or slopes. Subsidence can present problems which are costly to rectify. Slope can involve levelling, cut and fill of excavation, stepped foundations, drainage, etc. Fill may have to be transported to the site and it may have to be contained to prevent slip. Reclaimed, made, or infilled ground must be checked, as rubbish dumps, particularly those consisting of household refuse, may take many years to settle. It is also possible that the building will be constructed on an old site, say in a rebuilding programme in an inner-city area, and this could involve the diversion of existing services.

Before work proper can commence it is usually necessary to clear the site of obstacles, such as tree stumps, rubbish, boulders, old foundations, pavings, and of course the debris from any demolition work that may have been necessary. Bulldozers are useful for site clearing as they can travel over most types of ground. Small models termed 'calfdozers' are available for housing-estate, and similar work. Should the ground consist of grass or farm land — quite usual on sites for new towns or housing estates — it must be stripped of turf and topsoil, but this could be worth stacking for reuse later. Good turf is valuable and fertilized farm soil is often in demand for landscaping and making new gardens. In any case it is not possible to build on arable soil or soil which contains plant life. In fact the topsoil is usually quite unsuitable for building on. It may have become friable and weathered, or may contain decayed vegetable matter. In some instances it could be affected by sulphates or industrial waste, particularly where extensions to factories or workshops are contemplated. In normal conditions it is wise to assume that the top 150 to 250 mm of soil is unsuitable for building. This is a fairly normal procedure. The stripped subsoil should be firm and free from shrinkage.

The removal of the soil from the site may also have to be considered. Temporary site roads may be required to enable dumpers and trucks to move freely. Where the site layout permits, permanent roads are sometimes laid, complete with kerbs, before building begins, or the final coating laid after all building has been completed.

5.3 Types of excavation

Foundations for structures vary greatly. They can be deep or shallow, the depth depending not only on the type of foundation required but also on the nature of the site and the climatic conditions likely to be encountered. Substructure excavations vary according to the type of foundation needed. These will differ between small buildings, basements, sewers and drains, shafts, trial holes and pits, soakaways, rafts, etc. They may be dug in soils which vary in range and particle size from coarse stone to fine colloidal clay such as gravel, sand, clay, silt, sandy clay, chalk, or rock. Where narrow trenches are necessary, particularly when dug to considerable depths, some form of trench support may be necessary to prevent caving in. The bearing capacities and characteristics of soils and methods of timbering are considered in Level 2 of this series.

The foundation must be in ground which is sufficiently stable to carry the weight of the loadbearing walls. Some soils, particularly clays, can be affected by seasons; in winter or wet weather they are apt to swell, and shrink again on drying in the summer. It is necessary then that the trench bottoms in such conditions be made deep enough to remain unaffected by weather. Roots of some types of tree can also affect the moisture-content of the soil, which could lead to subsidence. Trenches, once dug, should not be left exposed for long

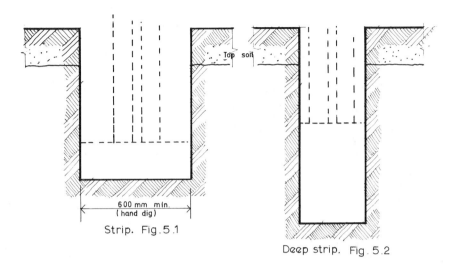

Strip. Fig. 5.1

Deep strip. Fig. 5.2

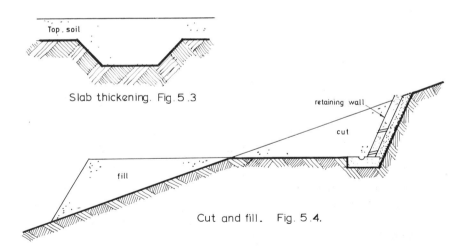

Slab thickening. Fig. 5.3

Cut and fill. Fig. 5.4.

Types of foundation excavation

periods as they are apt to shrink or swell after the concrete has been placed. Also trenches tend to disintegrate if left without support.

STRIP FOUNDATIONS. These are used mainly for small- to medium-scale building where traditional methods of construction are to be used. The trench is usually made not less than 600 mm wide to allow adequate working space inside the trench where hand labour is to be employed (Fig. 5.1).

DEEP STRIP FOUNDATIONS. These are used particularly in unstable soils, especially clay, where swelling and shrinkage are likely to occur. Where there are many identical foundations to be dug, as in the construction of housing estates, trenching machines may be used. The most usual method is to cut out trenches with a mechanical *backacter,* i.e. a tracked machine which operates a toothed shovel or bucket to scoop out soil with a backward action and deposit it in one movement. The bottom of the trench is then tamped and trimmed ready to receive concrete which is placed in the trench as soon as possible after digging (Fig. 5.2).

SLAB THICKENING. This is the usual method of providing a foundation under floors to support loadbearing partitions in small-scale buildings (Fig. 3.14). The excavation is normally about 150 mm deep and about 300 mm wide (Fig. 5.3).

CUT AND FILL. On sloping sites it is customary for ground excavation to be carried out to produce a level site. The amount of excavation cut from high ground usually balances the quantity of soil needed to create a level platform of appropriate size. Normally a bulldozer is employed for this purpose. The platform must be made large enough to extend beyond the structure. Often the high side of the site is supported by a retaining wall but the soil of the fill can be left to find its own angle of repose (Fig. 5.4).

STEPPED FOUNDATIONS. Sometimes excavations on sloping sites are stepped to keep the bottoms of the foundations level and so prevent sliding. The steps should always be in small increments of not more than, say, 300 mm to prevent undue settlement. A contractor would weigh up the economics of using a flat, deep strip trench, level all round and excavated by machine but requiring extra substructure construction, against the expense of creating steps by hand, especially on repetitive small-scale building (Fig. 5.5).

SURFACE SLABS. Combined ground and floor slabs with thickened edges are frequently used as rafts for buildings with walls of timber or other light material, especially when these are of a temporary nature. They are also used on unstable soils where subsidence is likely. The slabs are normally thickened at the edges in the manner shown in Fig. 5.6 not only to provide support but also to counter

Excavation work on construction sites | 45

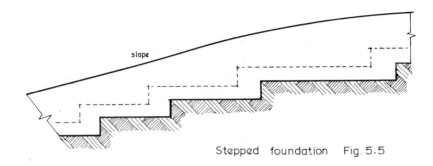

Stepped foundation Fig. 5.5

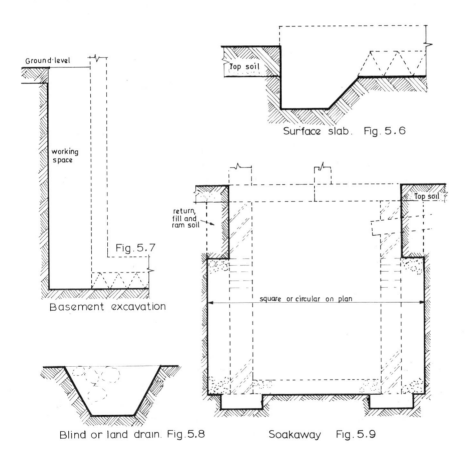

Surface slab. Fig. 5.6

Basement excavation Fig. 5.7

Blind or land drain. Fig. 5.8

Soakaway Fig. 5.9

Types of excavations

scour under the slab. It also overcomes shrinkage at the perimeter and serves to retain the soil under the floor and keep it stable.

BASEMENT AND RETAINING WALLS. Large excavations are almost always carried out by machine. The excavations usually extend to about 600 mm beyond the walls to provide a working space for timbering and external wall treatment (Fig. 5.7).

PAD FOUNDATIONS. Excavations for pads are usually carried with a backacter, as described for deep strip foundations, using a small bucket. Pads are used for both column foundations and as a substitute for strip foundations in some forms of building (Fig. 2.1). Narrow trenches only are required between pads to take edge beams which retain the soil under the slab floor.

PILING. Excavation for small-scale building substructure in unstable soils is normally carried out by mechanical auger to produce holes for short-bored piles. These are taken down to firm soil and can be placed very quickly in soft soils.

BOREHOLES AND TRIAL PITS. These are often dug on medium-scale sites in order to inspect the soil before building commences. For this type of building they are dug by backacter about one metre square and about 2.5 metres deep. They provide for soil examination and also show up any water problems which could affect building.

5.4 Removal of water

The most usual method of removing water from sites during construction is by means of ditches or trenches. Where possible these are dug to predetermined falls to drain the water out of harm's way. On high ground above sloping sites land or surface drains may be laid across the site to intercept water which could flow towards the building. These could be left in permanently if required as part of the contract (Fig. 5.8).

RETAINING WALLS. Where retaining walls are to be built they are sometimes provided with weepholes to allow water to seep through and be collected in open channels formed in the concrete area at the foot of the wall (Fig. 5.4). Should the wall and the channel form part of the contract, they are usually constructed first to provide a dry area on which to build.

REMOVAL OF WATER DURING CONSTRUCTION. This is frequently a problem in areas with a high water-table, particularly on sites where basements are needed. In such cases it is usual to dig a sump or well at the lowest part of the site and remove the water by pumping. Semi-rotary hand pumps are sometimes used for small work or intermittent pumping, otherwise mechanical submersible pumps are employed. Where permanent surface channels or road drainage are to be provided it is advisable to construct these as early as possible if this is permitted. Special diaphragm pumps are often used for dealing with muddy water and this may have to be treated before discharge into sewers. Care must also be taken in waterlogged areas when pumping to prevent undermining surrounding properties.

SOAKAWAYS. In some districts local authorities will not permit surface water to be drained into public sewers; in many cases soakaways have to be used, particularly in small-scale building. These are simply holes in the ground, either square or circular on plan and filled with hardcore or built with perforated walls and bottoms to allow water to drain away (Fig. 5.9). If these can be dug when the substructure and drainage system is being built, they may be used to drain the site.

FINALLY. On work of any importance great care must be taken not to overdig, i.e. excavate to levels lower than those specified, as soil once removed cannot be replaced; instead a weak concrete mix may have to be used to make up levels, which can prove expensive.

6 | Foundations

6.1 Function of a foundation

The primary function of a foundation is to spread loads applied to the supporting soil. The load itself can be a force acting on a structure or member and can be one of several types. A *dead load* is one represented solely by the weight of the walls, partitions, roofs, floors, and other permanent constructions including finishings, and an *imposed load* is one of people and equipment which functions other than as a dead load. This is sometimes referred to as a *superimposed* or *live load* though these terms are now deprecated.

A structure may also be called upon to resist *impact loads,* i.e. imposed loads whose effect is increased by sudden application, an occurrence frequently encountered in factories and industrial premises. A foundation may also have to withstand *wind loads* which can be exerted on the whole structure or part of it in varying degrees (Fig. 6.4). Wind loads will be discussed in Chapter 8.

TYPES OF FOUNDATION. The range and types of foundation are very wide and each has to be considered according to its purpose. Preformed foundations are now quite common, but the most usual method of making foundations is to mix the concrete to sufficient strength and to lay it down *in situ*. Foundations can also be of natural rock or firm soil which requires no preparation other than levelling up, but this is comparatively rare. The main types of foundation apart from strip, pad, and slab are the following.

Pile foundation: a construction formed of one or more concrete, steel, or wood supports driven or cast *in situ* into the ground (Fig. 6.6). Piles and piling will be dealt with at Level 3.

Raft foundation: a slab construction over the whole area covered by the building or structure (Fig. 6.2). These are used where the bearing power of the soil is weak or unreliable.

Sometimes, in order to spread the load at the foot of the wall or column, a *stepped footing* is used. This was commonly employed in earlier substructure, particularly under walls of heavy brickwork, but is not much used today.

6.2 Choice of foundation

In order to transmit the load satisfactorily from the building or structure to the soil, certain primary factors are involved. These depend on the function of the building and also on circumstances such as location and placing. The role of the foundation will be affected by the nature of the soil in which it is placed, the material of which the substructure is made, and the superstructure it will have to support. Building Regulations, Part D, 1976, covering structural stability, will be discussed at Level 2. Meanwhile, the fundamental considerations which affect the choice of foundation construction are as follows.

STABILITY. The substructure must not:

> break or twist under the load;
> disintegrate in the presence of water or sulphates in the soil;
> sag or crack because of the condition of the soil;
> slip when built on sloping ground;
> crack because of large roots;
> fail because of severe climatic conditions; or
> be undermined because of scour.

The final choice must be considered with some or all of the above factors in mind. It is necessary to know in what circumstances failure of small buildings could occur and these will now be dealt with briefly. Site investigation and exploration will be left until later.

LOADING. The dead, imposed, and other identifiable load should be calculated for each loadbearing wall or column and note taken of how these are to be transmitted through the structure to the foundation (Figs. 6.3 and 6.4). This is normally the task of the engineer, who is guided by the basic laws of mechanics and who must also observe the various building regulations, standards, and codes of practice. Wind and suction loads, for example, can present certain problems, and these will be outlined in Chapter 8. Dead-load forces, weights of materials, and similar information can be obtained from appropriate British Standards and official publications. As modern materials tend to weigh less than hitherto, and sizes of sections are now reduced to a calculated minimum; this results in considerable economy. Also site exploration now takes place on all but the smallest contracts, a factor which often results in savings of money and material.

SITE CONDITIONS. Local knowledge and existing ground records are often useful in helping to decide on the type of foundation needed. Mining activities, made ground, quarries, ditches, slow-moving slopes, ground faults, and weak and unstable soils can all affect the types of foundation needed. Evidence of flooding must be looked for; dry beds in the summer season could become waterlogged in winter, and ground water in underlying strata can sometimes be detected by digging trial pits. Subsidence can be dangerous unless special provision is made at the design stage. Seashore erosion can be a problem in some coastal areas. Local authorities, farmers, and residents can often give helpful advice on such matters.

The slope must be considered. Infilling can be expensive if transport is involved; soils may also have to be contained or treated to prevent slip; fill may have to extend well beyond the structure, and retaining walls could be needed at the higher levels. Creep or fracture could also occur at the foot of the slope downhill, but this is sometimes landscaped with binding grass or fibrous plants, which also reduces erosion.

Settlement of buildings is inevitable to some extent but foundation design should aim at keeping subsoil movement to a minimum. Provided that settlement is uniform over the whole area and is not excessive, the movement does little damage. If the amount of settlement varies at different points, however, the results can be harmful, and if this can be foreseen then foundation and substructure should be designed together to remain rigid enough to redistribute the load and prevent distortion.

Some buildings, particularly those of traditional unit construction of brick or block, are able to absorb a certain amount of shrinkage or distortion within the joints, which tends to make cracking less noticeable. Such differential movements, however, must be kept within limits as too great a dislocation could result in damage to the envelope and finishes. When foundations have been designed, their own dead weight should be added to obtain total bearing pressure on the soil.

6.3 Subsoil movement

Soil types usually consist of either rock, chalk, sand, gravel, clay, silt, mud, or combinations thereof. Some are excellent, though each can vary in itself. Rock, for instance, can weather badly, sand can be clayey or silty, clay can be either uniform, firm, stiff, or soft resulting in mud. Peat or infill is frequently useless as a bearing soil, also poor quality subsoils often result in movement, particularly in the presence of water.

Clays which shrink or swell according to the time of the year are common in some regions. This is invariably caused by change in moisture content due not only to rainfall but sometimes to the presence of roots, plants, grass, and other vegetation. Also a building on shallow foundations in some clays can be liable to seasonal movement as a result of rain, frost, drought, or root growth. Water from various sources such as streams, springs, high water-table, culverts, and poor soil drainage can also affect the subsoil.

Water in itself may not necessarily be harmful, as island buildings and bridge supports can testify. Small buildings, however, are often vulnerable if foundations have not been taken deep enough. Clearly any movement of the subsoil in the vicinity of the substructure can have disadvantageous effect varying from slight settlement to considerable subsidence. Long, dry spells have been known to cause widespread damage.

There are a number of ways of safeguarding foundations in clayey soils.

The most common are: diverting the water source by sumps or drainage; making the foundation deep enough; levelling and ramming the bottom of trenches; keeping the substructure clear of tree roots; and using piles or pads in unstable ground. Soil exploration and water containment before site operations commence could save trouble later.

6.4 Primary materials in foundation construction

Before the advent of Portland cement at the beginning of the nineteenth century, natural or primary materials were used exclusively in substructures. Good quality hardwoods were commonly used in underwater construction, especially those resistant to marine organisms and in foundations or piling in marshy soils. Natural stone or brick foundations have been in use for centuries, often laid dry, but frequently bedded and jointed in hydraulic lime mortar and with stepped footings to spread the load over a greater width (Fig. 6.8). In stone districts and for dyke building, foundations of dry stone walling are still used; the base course is laid in a well-rammed trench packed solid by use of small stone wedges or snecks inserted as the wall is built up. The great cathedrals built centuries ago bear witness to the efficacy of stone and hydraulic lime foundations.

Foundations today are almost invariably constructed of concrete. They consist of Portland cement, normal or rapid-hardening, fine aggregate or well graded coarse sand, and coarse aggregate consisting of natural gravel or crushed stone graded according to the designer's specification. Sometimes the aggregate is specified as 'all in', i.e. containing a proportion of material of all sizes as obtained from the pit, river, or other source. Concrete foundations requiring extra strength are sometimes reinforced with steel rods, but this and concrete production will be covered at Level 2.

Where the excavation bottom is muddy, it is sometimes necessary to put down a carpet of weak concrete in a fairly dry state before laying the foundations (Fig. 6.9). Its purpose is threefold: first, to keep the foundation concrete clean and free from impurities; secondly, to prevent seepage of liquid cement from the foundation while it is being laid; and thirdly, to bring the bottom of the foundation up to the correct level in the event of overdigging. The mixing of concrete which does not have specific strength requirements, such as that used for carpets for blinding or making up levels does not normally need supervising — as is the case with usual concrete construction.

Even when the bottom of the excavation appears satisfactory, it is usual practice to put down an underlay of polythene film before placing the concrete (Fig. 6.10). This prevents soil seepage into the concrete during pouring; it also prevents the concrete from being attacked by corrosion should the ground be acid; and prevents liquid cement from leaking into the soil while it is being poured. All primary materials used in concrete should be subject to scrutiny before the work commences; usually the contractor is required to submit samples of all materials to be used, and proper storage facilities should be provided on the site. These will be considered at Level 2.

Where concrete foundations are to be placed over uneven trench bottoms or on soils of unequal or dubious bearing capacity, it is usual to add m.s. bar or mesh reinforcement, particularly when bridging soft patches. It is sometimes added at vulnerable points such as junctions of floors and walls as well.

In order to prevent uneven settlement and cracking in concrete slabs, expansion joints are inserted at regular intervals. Horizontal m.s. bars are then placed across the joints at middle depth but inserted in sleeves to allow freedom of movement. In stepped foundations also, it is customary to insert mesh across the step to prevent separation at the junction. Before a concrete slab or raft is placed across a site it is almost always necessary to put down a bed of hardcore. This normally consists of clean broken brick, stone, or old concrete, tamped and rolled to a depth of about 200 mm (Fig. 6.12).

6.5 Pad, slab, and strip foundations

For small-scale building, simple pad, strip, beam, or slab foundations are normally used unless the ground is exceptionally poor or unstable, in which case rafts or short bored piles would be better.

PAD FOUNDATIONS. These are used where the load is transmitted to the ground from such points as columns and piers (Figs. 6.3 and 6.4). Lightly loaded panel walls and partitions can also be supported on edge-beams spanning from pad to pad, the latter designed to take the whole load (Fig. 6.5). In some forms of system building, panels of standard storey height are designed to be entirely self-supporting with vertical edges grooved and spaced to allow for columns *in situ,* formed to rest directly on the edge-beams centrally over the pads themselves. The advantage of this is that, apart from pad bearing, the edge-beam carries no load but acts only as a retaining wall to hold the subsoil and hardcore in place beneath the floor. This enables the pads to be constructed at any depth to suit the ground slope and the nature of the soil. An isometric sketch of this arrangement is given in Fig. 6.5. Pad foundations are also used extensively for lightweight or temporary building where short brick or block piers are built up from the pads to support the superstructure (Fig. 6.11). A similar arrangement is shown in Fig. 6.3b but this uses a strip foundation.

STRIP FOUNDATION. This is commonly used for loadbearing structures of concrete, brick, block masonry, or stone. It is almost always of concrete and constructed as shown in Figs. 5.1, 5.2 and 6.1. Hand digging is customary for this type of trench (Fig. 5.1), but machine excavation for the type shown in Fig. 5.2. The former has to be made wide enough for the bricklayers to work in, otherwise such a width would not be necessary.

Foundations | 53

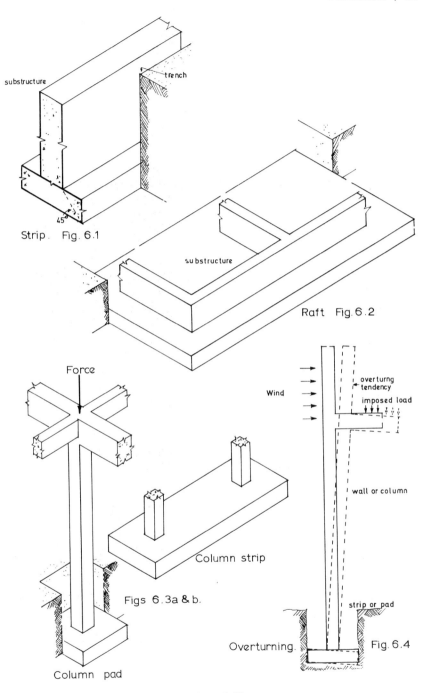

Strip. Fig. 6.1

Raft Fig. 6.2

Figs 6.3a & b.

Column pad

Column strip

Overturning. Fig. 6.4

Types of foundation

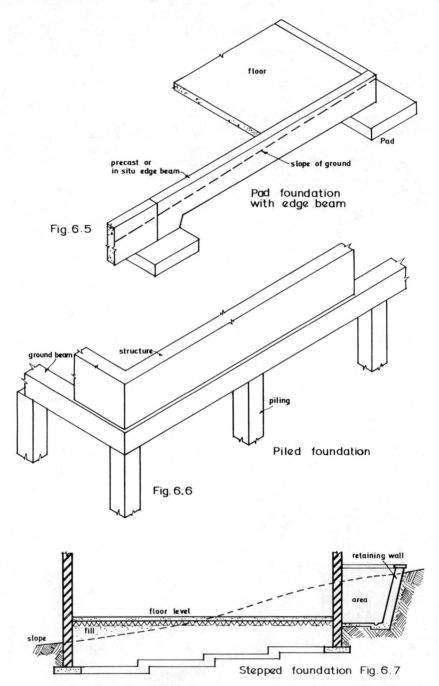

Types of foundation (cont'd)

In spite of various theoretical width calculations for different loadings and soil capability, the contractor may find it profitable to use the same trench-width throughout, using a size which will meet the worst conditions. Trench excavation by backacter with a standard bucket is usually quicker and more economical, despite any extra concrete material that may be necessary. This method can also be used to eliminate stepped foundations where the slope of the ground is not too severe. He would then put in a level trench-bottom all round.

The depth of the foundation below ground-level will depend on three conditions: first, the need for adequate soil bearing, secondly, the need for protection against frost, and thirdly, the need to avoid disturbance by movement in clay soil, tree roots, vibration, subsidence, etc. Frost can penetrate up to 600 mm in some areas and this is now taken as a reasonable minimum, though in clay, one metre is recommended. Further information may be had from B.R.E. Digests Nos. 63, 64, and 67.

The depth of concrete in strip foundations should not be less than the extension beyond the face of the wall (Fig. 6.1). In any case it should not be less than 150 mm. Soft patches of ground should be spanned by introducing reinforcement into the foundation concrete.

RAFT FOUNDATION. This sort of foundation is used where the whole building area is required to be covered, and is usually in the form of a concrete slab. A raft foundation is needed where the bearing pressure of the ground is very low, as with infilling, made ground, or soft ground. The raft must be of sufficient area to avoid overstressing the soil, and stout enough to carry the load placed upon it. It may be laid at some convenient depth below ground (Fig. 6.2), or it may be placed at ground level as a surface raft (Fig. 6.12). Such rafts combined with edge-beams as shown are frequently used for timber-framed or prefabricated building and also for light structures where subsidence or other forms of ground instability is likely.

Normally, rafts used in solid floor construction are about 150 mm thick. In temporary work such as site offices, sports pavilions, etc., the thickened slab edge is not usually reinforced, though this is desirable for more permanent structures. Where such slabs are used on partly made-up ground such as cut-and-fill sites (Fig. 6.7), they are often reinforced over the infilled area. This will be dealt with further at Level 2.

6.6 Foundations to beds and pavements

The provision of hard surfaces on areas surrounding buildings is a functional as well as an aesthetic requirement. In urban areas people demand suitable surfaces, particularly in precincts, pedestrian areas, parks, and footpaths and similar walkways. The design of recreational, shopping, industrial, and other

56 | Construction Technology

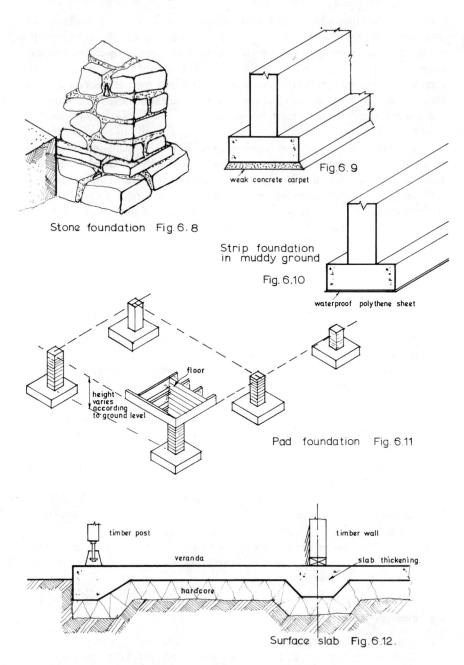

Stone foundation Fig. 6.8

Fig. 6.9
weak concrete carpet

Strip foundation in muddy ground
Fig. 6.10
waterproof polythene sheet

floor
height varies according to ground level

Pad foundation Fig. 6.11

timber post
veranda
timber wall
slab thickening
hardcore

Surface slab Fig. 6.12.

Types of foundation. (cont'd)

centres usually rests with the architect or landscape planner whose function it is to combine the design of hard surfaces with soft landscaping and the use of street furniture, parking signs, and recreational equipment.

Roads for vehicular traffic, car parks, hard standings, estate routes, and turning areas are now normally referred to as *pavements* by road engineers. In general, however, a pavement refers to a pedestrian sidewalk to a road.

CHOICE OF MATERIAL. This will depend on the architect or landscape designer and the need to separate pedestrian from vehicular traffic. There may be a dual need when light traffic also uses the circulation area, or when public utility vehicles or municipal appliances have to be accommodated. In such cases it may

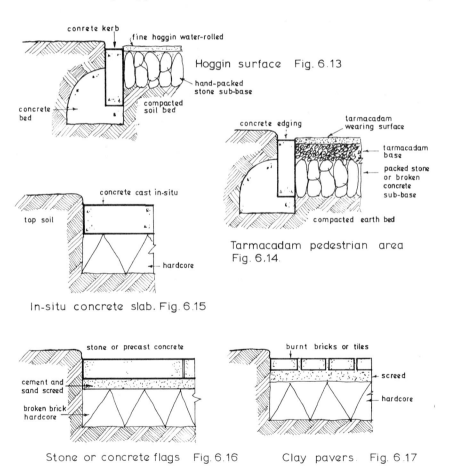

Foundations to beds and pavements

be necessary to include a road base or sub-base to withstand the load (Fig. 6.14). Alternatively, the foundation could be increased to 150 mm thick and a flexible top course of paving laid on a base course of 1:3:6 concrete 100 mm thick (Fig. 6.15).

TYPES OF HARD SURFACE. The choice of pavement depends very much on the purpose for which it is intended. A cheap and widely used surface for footpaths and pedestrian areas is that of stabilized *hoggin* or gravel (Fig. 6.13). It is attractive when well kept, but needs regular attention. The sub-base needs to be treated with weed-killer, and some form of lateral retention at the edges is essential. The surface may be stabilized with the addition of bituminous emulsion (or Portland cement if the soil is fairly sandy) after which fine gravel is rolled into the surface.

Tarmacadam. This is a graded stone mixture coated with tar, spread on a prepared foundation, and rolled until compacted (Fig. 6.14). Various stone aggregates can be used according to the sources of supply. The thickness will depend on surface requirements, which are quite numerous and which are described in the appropriate British Standards.

In situ *concrete* is very popular for hard surfaces as it is cheap and can be textured by exposing the aggregate or by using coloured materials (Fig. 6.15). It can also be laid in bays alternating with pre-cast pavers, cobbles, or other deterrent paving. It may also be panelled by framing into bays formed by strips of baked clay paving bricks usually about three courses wide. The concrete bays are usually float-finished.

Pre-cast concrete flags. These are made of finely crushed stone and Portland cement (Fig. 6.16). They can be obtained in a wide range of sizes, patterns, and colours and are also available with textural surfaces if required. For pedestrian use in patios, footpaths, etc., they are usually 50 mm thick; where light traffic is expected the thickness should be increased to 65 mm.

Clay tiles and brick pavers. These are also in wide use (Fig. 6.17). Pavers are dense and frost resisting and can withstand water and quite rough treatment. Quarry tiles are cheaper but thinner and not so durable. All can be laid in attractive patterns such as herring-bone, basket-weave, etc.; concrete block paving is also in wide use for lightly trafficked roads and paved areas; other pavings will be covered at Level 2.

Surface water run-off. Hard surfaces should be laid to a fall of 1:120 and collection effected by gullies, channels, and other means. The basic principles of drainage are dealt with in Chapter 16.

C. Superstructure

7 | Function of basic structures

7.1 Relationship of superstructure to substructure

As described in Chapter 4, a structure is considered as consisting of a substructure and a superstructure, the former representing that part which is situated below ground and the latter that part which is above ground; the design and function of the structure determines the relationship between the two. For example, a steel lattice (Fig. 7.1) or a concrete-framed structure designed on an open plan with as few intermediate columns as possible would need quite a different substructure from that of a building of cross-walled construction (Fig. 7.5). It may also be required that a substructure should absorb the effect of sound waves, vibration, damp, temperature variation, and movement, particularly when subject to seasonal change.

The superstructure must also relate to its support. It can be subjected to a variety of stresses and strains through dead-, imposed-, impact-, wind-, or suction-loads, each of which must be absorbed by the substructure (Fig. 7.2(a)). Depending on site and soil conditions, the substructure can be made either flexible or rigid, and is sometimes required to hold down as well as hold up the structure (Fig. 7.2(d)). This may also occur in buildings subject to deformation from causes other than side thrust (Fig. 7.3), such as waterlogged soil. The design of a substructure, other than that of a simple building, is usually governed by both the subsoil and the superstructure itself. The slope of the ground often plays an important part in the relationship of the substructure to the superstructure.

7.2 Basic concept of a structure

In Chapter 6, the forces acting on a structure or member were described as consisting mainly of two types — dead loads and imposed loads, both transmitted from the point of application through the structure to the point of support. Apart from wind loads, the transfer of forces from one element to another is mainly vertical. We shall now consider a few examples.

A flat roof or floor (Fig. 7.1(a)), can transfer its load directly to the walls, which in turn transmit it to a strip or raft foundation. The building can also be of cross-wall construction (Fig. 7.2(a)), i.e. made up of horizontal slabs supported by walls, both internal and external, which transfer the load to the strip or raft foundation. Should the ground alone be unable to carry the weight, piling may be used as shown in Fig. 7.2(b)).

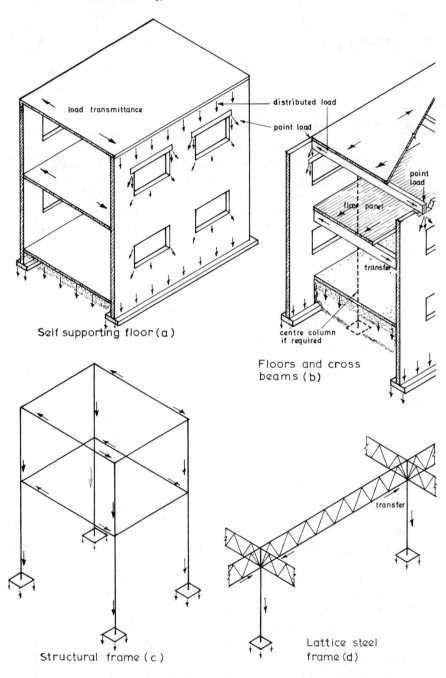

Transmission of forces Fig. 7.1

Function of basic structures | 63

Should the width of the building between the walls be too great for the floor to span unsupported, then it is usual to insert cross-beams as shown in Fig. 7.1(b)). The floor loads are then transferred to the beams, which in turn transfer them to the walls. It will be seen from this arrangement that the floor or roof load is spread evenly over the supporting beams; this is known as a *distributed load.* These beams, however, are supported by the external walls as *point loads,* and the wall must be strong enough at this point to support this concentrated load. Sometimes intermediate columns are necessary to carry the cross-beams as shown.

In modern construction it is usual to dispense with traditional external loadbearing walls and to use the structural frame instead (Fig. 7.1(c)). In this case the floor and roof loads are transferred to the main beams which in turn transmit them to columns and so to the pad foundations. It will be seen that the external beams do not carry floor loads but support only the external 'envelope' and also help to stiffen the structure. Frequently the external walling consists of lightweight cladding or *curtain walling,* i.e. a non-loadbearing wall constructed outside and continuously over the structural frame to enclose the

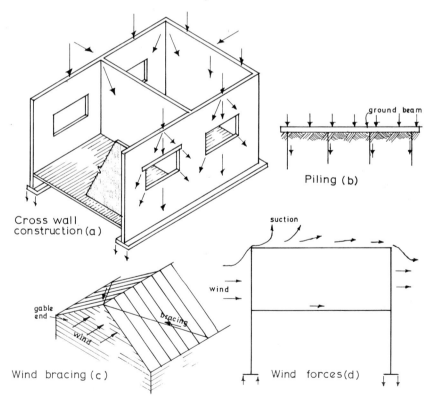

Transfer of loads Fig. 7.2

building or structure (BS 3589). Where large covered areas are required such as garages, factories, or transport depots, the number of intermediate columns are reduced to a minimum by means of lattice girders (Fig. 7.1(d)).

Wind loads. Forces acting on structures are not always vertical. Wind plays an important part in structural design and must be carefully calculated (*see* Chapter 8). Wind blowing horizontally with sufficient force tends to increase foundation pressure on one side of the structure and reduce it on the other (Figs. 7.2(d) and 7.3). It is common practice to introduce some form of bracing to resist wind force and to stiffen the structure (Fig. 7.2(c)). In cross-wall construction of traditional brick or blockwork, the walls themselves provide resistance (Fig. 7.2(a)). When dealing with wind forces, the suction acting on flat or low-pitched roofs must not be ignored and some form of holding-down device may have to be provided (Fig. 7.2(d)).

Other forces. In addition to those outlined above, other forces can also act on beams, columns, and foundations. Figure 7.3 gives a few examples: (a) shows a load simply supported on a column at a central point and the force in the column is vertical; in (e) the beam is firmly fixed to the column, which extends upwards to carry the floor above (in this case the beam tends to bend the column thereby creating uneven pressure on the foundation); in (b) a cantilevered action is shown whereby a balcony beam projecting from a column or wall also tends to produce a turning effect on the structure; and in (d) the thrust of the roof rafters tends to force the walls outwards. Other examples could also be given.

7.3 Basic types of structure

The basic purpose of a structure is primarily that of shelter. The environmental 'envelope' as we now understand it was developed mainly from primitive forms which varied according to location and function, and were built according to local needs with whatever materials were to hand. Nomadic peoples preferred tents, but settled communities favoured permanent structures built as sturdily as climate demanded and materials permitted.

Permanent structures, however, varied greatly according to climate and materials available. In hot, dry climates where the sun was fierce, nights sometimes cold, sand storms not infrequent, and woody plants scarce, building developed differently from the way it did in warm, humid zones where vegetation was lush, water more plentiful, temperature generally lower, and shelter less important than cross-ventilation. Polar regions dictated other requirements (Fig. 7.4). In all cases, however, a microclimate was created to make living conditions tolerable.

As countries progressed, however, imported materials became more widely used and the styles of building changed. Priority is now often given to the function of the structure rather than to its situation, and building technology is required to keep pace with standards demanded for modern comfort and

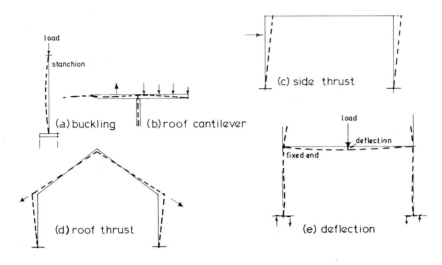

Deformations Fig. 7.3

essential services. Today many structures are not built for human occupation, for example storage or automatic processing plants, but these too often need carefully controlled microclimates. However, only the functions of typical structural forms will be dealt with here and those which have now evolved are of three basic types, namely *skeletal, solid,* and *surface.*

Skeletal. This is basically a framework arranged in such a way as to resist forces natural to the environment. The skeletal frame is made secure to a foundation and enveloped by a skin or membrane of the most suitable material available. The primitive lean-to (Fig. 7.4(a)), later developed into a self-supporting structure such as the tepee or wigwam (Fig. 7.4(g)), which is truly skeletal. In Africa, a square or round hut with walls of woven split bamboo or wattle mats tied to an upright framework of poles and with a gabled or conical roof of woody stalks covered with thatch, could also be classed as skeletal. Broadly speaking, the skeletal structure is a frame supporting a non-loadbearing envelope or fabric. Thus the modern structural lattice and triangulated frames of Figs 7.1 and 7.6 are classed as skeletal. Another popular form is that of a grid dome of triangulated framework, usually partly spherical, to which a covering is applied (Fig. 7.6). These need no intermediate support and can be made to great size.

Solid. In settled or pastoral communities with a need for permanent building, the solid structure was evolved consisting basically of sturdy loadbearing walls to support the roof and resist the elements. The earliest form was probably the portal (Fig. 7.4(b)). A solid structure is generally one in which all loads are transferred to the foundation through the walls themselves. An early example of this is shown in Fig. 7.4(c) where wet earth is kneaded and mixed with grass to

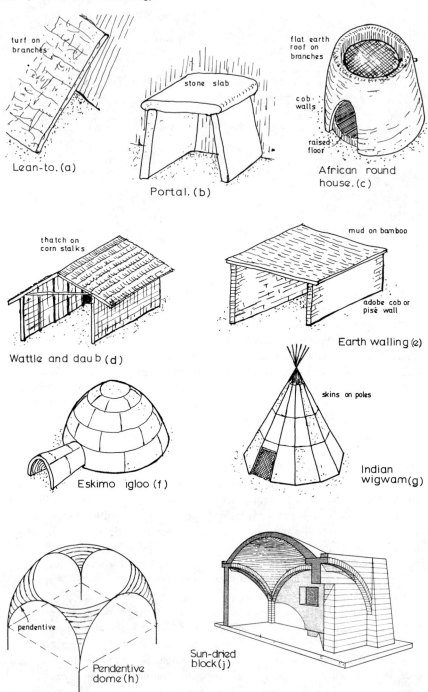

Basic types of structure   Fig. 7.4

form cob, still used today. From such early forms developed the traditional, self-supporting structure, as shown in Figs. 7.1(a) and (b).

The distinction between skeletal and solid can sometimes become blurred. This can occur in primitive building where daub is applied to wattle and in modern structures where the envelope or cladding is stiffened in some way. Between the wars it was customary for a public or commercial building to be constructed as a heavy mild steel frame and then encased in solid stone or brickwork. A usual type of solid building today is that of cross-wall construction (Fig. 7.2).

Surface structures are those in which the covering or envelope, either wall or roof, can act as a stressed skin, either in tension or compression, to hold the structure in place. The oldest form of tensile surface structure is probably the tent, where woven cloth or other fabric is held up by poles and stretched tight by guys secured to the ground, the stability of the structure being supplied by the material. This is a method much in vogue today, where large areas of woven mesh or fabric draped over poles, pylons, ribs, and braced framework, are held taut to create some quite outstanding shapes. Such covers can also be suspended from the external frame, a system which is also currently in fashion.

The most widely used form of surface structure with the 'envelope' in compression is the shell. Primitive 'envelopes' of this kind are still to be found in some parts of Africa in the form of basket-weave of split bamboo or stalks. These are constructed as an inverted bowl fixed to the ground, then daubed with mud and left to dry. This provides the necessary rigidity to hold the casing firm.

Probably the earliest type of doubly-curved shell is that of the dome, of which the Pantheon in Rome is an example. Early cathedral builders also favoured construction of self-supporting curved surfaces such as domes or vaulting usually built of stone (Figs. 7.4, 7.5, and 7.6). By the use of modern materials of carefully calculated thickness, surface coverings can now be constructed as very thin shells designed to cover large areas without intermediate support. A further geometrical development is that of the hypobolic paraboloid (Figs. 7.5(h) and (j): in this case the skin is in compression in one direction (B), and in tension in the other (A)).

7.4 Typical structural forms

The shape and position of a building component in a structure can often affect its strength and so result in considerable saving. For example, the rigidity of a corrugated sheet (Fig. 7.5(d)) is obviously greater than that of a flat one. A plank of wood laid flat to span an opening will bend much more easily than if it were placed on edge; in this latter position, however, it could twist or whip unless strutted in some way (see Chapter 13).

Similarly, a tube or cylinder laid horizontally across a span would support a greater weight than if it were first flattened and then loaded. A vertical tube can

68 | Construction Technology

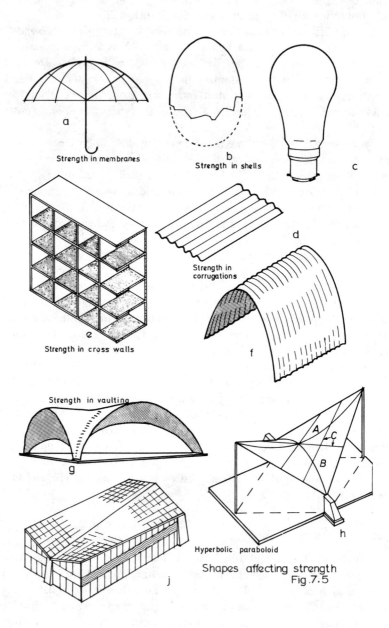

Shapes affecting strength
Fig. 7.5

also be compared with a solid rod in the same way. Should they both be of the same length, contain the same kind and amount of metal and weigh the same, the cylinder would carry the greater load, as the rod would tend to buckle earlier. There is a limit, however, to the size of the cylinder; it would tend to collapse if the wall was too thin. Figs. 7.5(a) to (c) give examples of well known objects — an umbrella, an eggshell, and an electric light bulb — where shape has been utilized to provide the necessary strength.

The main object of a structural designer is to obtain maximum strength or coverage with minimum material or, as one famous dome designer puts it, 'to do more with less'. This concept has resulted in new shapes and methods of construction already outlined. The materials used would depend on whether the structure is skeletal, solid, or surface, and the decision would be affected by questions of labour, material, and function. The form may also be influenced by economical, aesthetic, environmental, experimental, or even political considerations.

With the possible exception of reinforced concrete, heavy or bulky materials are now used less frequently (Fig. 7.6). With cheap and speedy forms of transport available, some rural areas which up to now have used indigenous materials are tending to lose their local character. Economy and function are often the main considerations, apart from stringent requirements regarding appearance, safety, ease of erection, reproduction, insulation, weight on foundations, and general durability. These now often take precedence over traditional image.

The manner in which materials are combined can also influence structural form. Concrete, for example, can be produced in a number of ways to produce various textures, strengths, or shapes according to whether it is cast in place, preformed, or added as a surface skin. Standardization and repetition in production have also had an influence on form. This also applies to materials such as timber, now processed as a wide variety of timber-based boards. Scarcity of labour has also affected modes of production and styles of building, and played a large part in industrialization and method. Plastering, brickwork, and masonry, classed as 'wet trades', and much work *in situ* have now been superseded by dry construction, usually preformed, which has resulted in new systems of building.

7.5 Component parts of a structure

Building construction is the design of a fabric and the method of putting it together. The fabric may be dependent on the framework or the system of connected parts. But building as a technology has changed rapidly during the present century; it is no longer mainly a craft-based trade where units of construction are shaped by hand and manhandled into position.

Each main part of a building, or *element* of construction, has its own functional identity, such as foundation, floor, roof, or wall. The term *framework* as used above should not be confused with *framed building* (Fig. 7.1(c)), which

70 | *Construction Technology*

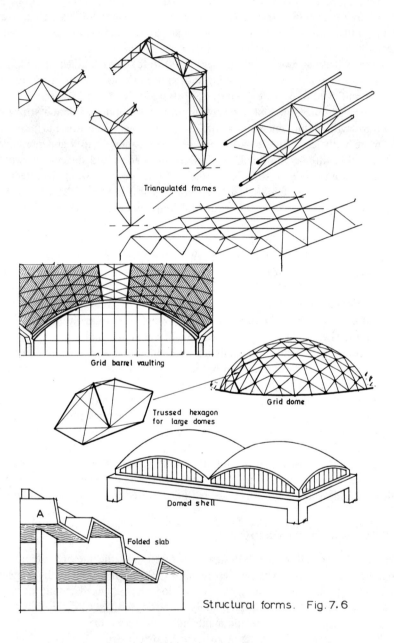

Structural forms. Fig. 7.6

relies on a frame rather than on loadbearing walls for its strength. The structure and fabric of most buildings are made up of components consisting of *sections, units,* and *compound units* (see Chapter 1). The latter may be identified as composite articles each complete in itself but which is intended to be part of a complete building, such as a door, door-frame, window, or sink unit.

The term *section* is widely used in building parlance and refers to materials formed to a definite cross-section but of unspecified length. Sections are normally produced by a manufacturing process such as rolling, drawing, extruding, or machining.

Rolling is performed either on cold metal sheet by passing it through flat rollers or by forcing hot metal billets between rollers shaped to the required profile to produce structural sections. Sheet or plate metal can also be cold-rolled into a wide variety of shapes including rectangular hollow sections (R.H.S.), door- and window-frames, light standard structural sections, etc. Hot-rolled sections are usually in the form of m.s. sections, angles, channels, tees, and other shapes.

Drawing is the method used to produce wire, cable, and thin bars. This is done by pulling metal through small apertures of required shape and size.

Extrusion is performed by forcing plastic material through an aperture to produce a given section.

Machining is the automatic cutting or finishing process applied to any material, mainly metal or timber, to produce a desired result.

Modern building technology makes great use of compound units or combinations of sections and units, such as glass bedded in window-frames to give a complete window. They can also be a combination of different shapes of the same material such as rolled-steel sections and metal sheet to form panels or stanchions. *Compound unit components,* as the name implies, are assemblies such as built-up panels of timber or other framing containing complete built-in windows or doors.

Whereas small units such as bricks or blocks are normally placed by hand on site, the practice is growing of fabricating these into compound unit components either on- or off-site, and then lifting them into position by crane. This is generally cleaner and quicker than working *in situ* and is usually more accurate. However, there must be a need for a sufficient quantity of each unit to justify the initial outlay. The process always involves dimensional co-ordination, a system whereby all manufacturers concerned with supplying the articles which make up the finished components, work to a predetermined set of dimensions and tolerances.

Many large organizations now produce finished buildings complete. These are transported to the site to be placed on prepared foundations and are often ready for immediate occupation; other products may be of individual k.d. elements, ready to be assembled on site, while some can provide the external envelope only, leaving the interior finishing to be completed by the contractor. A wide range of systems is available, each with its own characteristics. These will be covered in more detail at a later stage.

8 | Functions of the 'external envelope'

8.1 The 'external envelope' and its elements

The composition and construction of the building fabric comprising the 'external envelope' is governed by both the external environment and the internal environment needed in the enclosed space. In small- to medium-scale structures the envelope normally consists of walls and roof, each element being of traditional construction and capable of withstanding external conditions of stated limits of severity. These elements embrace whatever components are necessary to meet specific building requirements in order to provide satisfactory conditions for human comfort.

In larger or more complex structures there may be special needs such as security (strong rooms), thermal insulation (warm and cold rooms), light control (dark rooms), clean air (hospitals, laboratories), or temperature control (air conditioned premises). The envelope could then be of newer materials and form, designed to modify the interior environment and create conditions to satisfy higher functional standards.

8.2 Primary functions of the 'external envelope'

The importance of the requirements outlined above will vary according to the conditions needed in terms of warmth, light, ventilation, safety and security, latitude, and location. With these must also be included the combined effects of climate. Such considerations must be designed to meet whatever building regulations, by-laws, or specific standards may be stipulated. Some of these have been outlined in Chapter 7. Those of the 'external envelope' may be identified as follows:

strength and stability,
weather exclusion,
thermal insulation,
sound insulation,
durability,
fire-resistance, and
feasibility.

STRENGTH AND STABILITY. Owing to the development of a scientific and mathematical approach to building construction, the structural requirements of a building in respect of strength and stability have become more clearly identified and can now be quite accurately defined. But stability has also resulted from constructional techniques which have been developed by trial and error since the earliest days of building. The primary function of a structure therefore, is to resist:

structural collapse;
structural dislocation;
local damage, i.e. settlement, cracking; and
ground movement.

It is general practice for the engineer of a new project to submit structural designs and calculations to conform to building regulations and statutory requirements. Unfortunately such conditions, though essential in the main, tend to discourage innovation, especially when structural failure involves heavy financial and social penalties. But methods of design are changing. The *factor of safety* method of determining working stresses of materials is giving way to the *load factor* as derived from large-scale testing of steel and reinforced concrete structures during the past five or six decades. Other methods are also in use to take account of local conditions.

WEATHER CONDITIONS. One of the basic reasons for the erection of shelters is as a protection from the weather, mainly rain, though wind and snow must also be considered. Exclusion of weather by the 'external envelope' is almost always identified with dampness, whatever its cause. The frequency and intensity of rainfall is well recorded by weather stations everywhere and despite seasonal variations the designer normally knows what to look for. The Building Research Establishment has developed an index of exposure to driving rain, which has been obtained by combining the average rainfall with the average wind speed for a given locality; this in itself can be affected by the severity of the exposure or the height of the building.

There are several methods of resisting rain penetration in common use. The first is to provide a waterproof skin over the 'envelope' either in the form of cladding, curtain walling, or external rendering to walls, and an impervious covering to the roof. The second is to make use of semi-absorbent material whereby moisture is held in the body of the wall until weather conditions permit it to evaporate. A third is to use walling containing interstices which prevent capillary action through the wall from taking place. A fourth and most popular method is to provide a discontinuous wall by the insertion of a cavity or an air gap (Fig. 8.2(d)). This breaks up capillarity, permitting water to evaporate or drain off inside the cavity. The inside skin may also be of timber-framed construction with an impervious lining to the inner face of the cavity.

THERMAL INSULATION. Among the primary functions of the 'external envelope' is protection from extremes of temperature, either heat or cold. Standards of

Functions of the 'external envelope' | 75

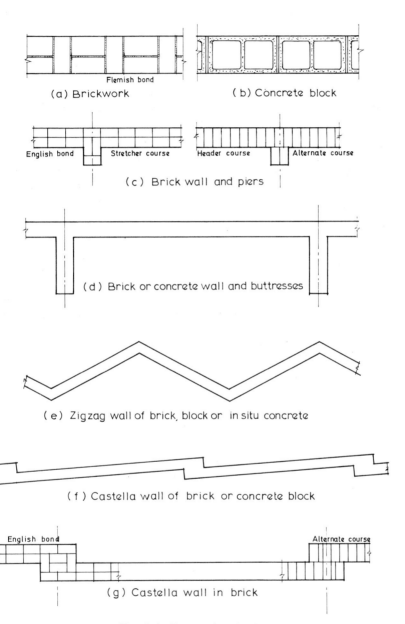

Fig. 8.1 External wall plans

human comfort have risen during the last few decades and building regulations now demand a calculated measure of thermal insulation for domestic and industrial users. It is now necessary to have some understanding of the insulating qualities of building materials and fabric and to know how best to use them.

In designing for thermal comfort for human beings and for suitable conditions for industrial processes and storage, four characteristics of the environment have to be considered: temperature, humidity, air movement, and heat flow between occupants and the enveloping structure. The design of the building is concerned mainly with temperature and flow of heat or *thermal transmission*.

Environmental considerations. Air temperature which affects heat gain or loss through conduction and convection is usually measured by a thermometer, though there are other considerations which need not concern us here. *Humidity* is affected by the rate of evaporation in the atmosphere, which directly controls the rate of heat loss. Relative humidity, expressed as a percentage, is dependent on the dry-bulb temperature and the concentration of moisture in the gaseous mixture of water and air. Air velocity is usually measured by a Kata thermometer.

The rate of heat-flow through a material is dependent on its density and thermal conductivity, and standard values of these may be had from reference data published by the Building Research Establishment. It is essential to understand something of the thermal properties of materials for different constructions. Cavities, for instance, can play an important part in thermal control.

NOISE AND SOUND-INSULATION. Sound-insulation qualities are usually measured by the grading of the materials. Grade curves for impact-noise and airborne-sound insulation are obtainable from a number of publications of the B.R.E. Noise levels in decibels may be measured from tests with impact machines.

Lightweight concrete is one of the newer forms of construction used to combat noise and sound-penetration (Fig. 8.2(d)). Normally the most influential factor is the weight of the structure, but, with new developments in industrialized building, sound-insulation factors other than weight have to be explored. Double-leaf walls provide insulation in excess of their weight contribution, but, to be effective, air gaps should be closed to prevent sound movement. This can sometimes prove difficult with the dry construction now popular in concrete and timber-framed buildings.

Roofs also need special attention. Pitched roofs are seldom airtight, but flat roofs can generally cope with normal noise, depending on their construction. It should be possible to provide for the effect of noise before the building is erected, by measuring the existing exposure to noise, but allowance has to be made for the increase in sound resulting from pressure at the face of the new building after it is put up. Where new roads or highways are to be built near the structure, provision must be made for the gradient of the road, ground absorption, and traffic frequency; and soft grass shoulders or noise screens should be provided if necessary.

DURABILITY. Durability may be defined as the continuing ability to meet the functional requirement of the building element. Functional requirements in themselves must be considered in terms of initial cost, upkeep, maintenance charges, and replacement outlay. The client must decide on the standards required and what he is prepared to pay for them.

Deterioration of the building takes many forms. Initially it may be superficial and affect only the appearance, but, if neglected, can lead to failure involving major repair. Regular maintenance is usually essential; this could take the form of renewed paint coatings, jointing materials, mortar pointing, cleaning down, etc., to reduce the effect of atmospheric pollution. Sometimes periodic maintenance can be reduced by careful detailing to prevent staining and streaking due to deposits and water run-off from overhanging ledges.

Water. The most common cause of deterioration of the building element is the presence of water. This may be due to frost action, chemical reaction, crystallization, or volumetric change. Frost action can be due to the expansion of ice crystals exerting pressure in pores or cracks in the surface of materials; chemical action can also cause deterioration in ferrous metals which are particularly susceptible to corrosion. Electrolytic action can also cause metallic corrosion, the most common example of this being the combination of copper and steel in water-supply systems. Water which has been in contact with Portland cement or mortar can also cause corrosion in lead or aluminium.

Acids. Acid solution, caused by plaster containing calcium sulphate coming into contact with certain timbers such as oak, can corrode most metals. Corrosion of exposed steel reinforcement in r.c. work can also cause failure, particularly where cracking takes place in loaded members, or where poor quality concrete has been used. Breakdown of building materials could also be due to atmospheric acids; for instance, there have been cases where Portland cement has been attacked by sulphates, which has ultimately led to the failure of the structure. For this reason, sulphate-resisting cement can be specified if circumstances permit. Brickwork can also contain soluble sulphates resulting in loss of strength and stability.

Crystallization damage in porous materials such as brick or stone can also take place where materials absorb water containing substances which then crystallize on exposed surfaces. Most soluble substances cause little damage but do create an unsightly patchy appearance. In some instances, however, damage can result from crystallization.

Movement. Volume change due to moisture-content does occur in many woody, clay, or concrete products, and failure can result where large-scale expansion or contraction takes place. Local cracking can also be induced by the restraint in the structure. Freedom of movement by use of expansion joints or similar means must always be considered.

Timber and timber-based products can cause problems due to movement during seasoning, humidity variation, and central heating. This can be over-

78 | Construction Technology

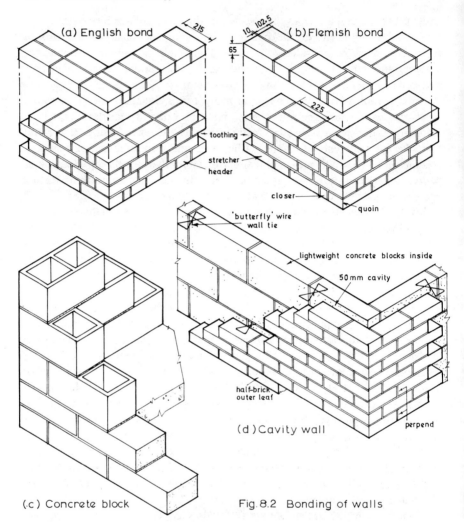

Fig. 8.2 Bonding of walls

come to some extent by changing the direction of the grain where possible, as in plywood, or by using jointing devices designed to permit expansion. Other causes of deterioration are the effect of sunlight on building materials — due mainly to ultraviolet light — and also to biological attack through fungal and mould growth. These will be covered at another level.

Protection. Thin surface-coatings are the usual means of protecting timber and mild steel. These almost always consist of synthetic polymer materials which are usually superior to natural oils and resins both in terms of cost and improved performance. Corrosion of mild steel can be countered by the use of thin zinc coatings frequently incorporated during manufacture by such processes as galvanizing and sheradizing. Thicker bitumen-based coatings are commonly used for heavy duty protection.

FIRE-RESISTANCE. Fire-resistance is mainly concerned with the choice of materials and structural method. The duration of a fire and the temperature reached are the critical factors in determining the fire-resistance required in building elements if failure is to be avoided. An essential need in structural design is to provide for walls and floors to act as highly resistant fire-barriers which not only contain the fire within a given area, but also permit safe escape-routes for the occupants.

The period of fire-resistance is stipulated in fire regulations governing the provision of, and means of escape. The fire-resistance of building elements must also be considered in terms of non-combustibility, ignitability, fire propagation, and surface spread of flame, all of which are covered in the appropriate British Standards.

APPEARANCE. The appearance of the finished structure is a vital factor to both the designer and the client himself. In the eyes of the latter, it frequently takes precedence over the durability of the structure, about which he may know very little. Thus, whatever the qualities of the structure may be, appearance must not be underrated. The finished effect of traditional materials such as brick or stone can usually be predetermined, as can their appearance after a period of weathering. The latter will depend on the location and its atmosphere and this is normally a first consideration at the design stage. Atmospheric pollution, weather and external exposure, wetting, drying, and abrasion are essential considerations.

External exposure of samples of wall panels or trays over as long a period as possible gives a good indication of the finished appearance, which can be helpful where matching or contrast with the surroundings is desirable. Examination of existing buildings in the vicinity constructed from similar materials can be a useful guide. The question of appearance is closely connected with durability, stability, and weather exclusion, each of which should take priority over visual effect of the finished product. Careful detailing in design can also play a vital role in the final appearance of the structure.

FEASIBILITY. The feasibility of a project can be assessed by the fitness of the design for its intended purpose, and this is almost always the responsibility of the designer acting on the instructions of the client. The effectiveness of the structure as a whole can be gauged from the application of the primary functions outlined above. There are also other considerations including building regulations, by-laws, and acts of parliament to be taken into account. Further constraints would apply in matters of cost benefit, value analysis, and similar factors outside the scope of this level.

What should be considered at an early stage is the feasibility of the building process both during the pre-construction and post-construction period. This covers the manufacture of components, transport, access, loading and unloading,

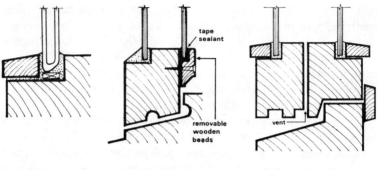

Typical factory-sealed double glazing units   Typical glazed in situ double glazing   Typical double windows, coupled type

Fig. 8.3

storage, casting, assembly, erection, fixing, protection, and workmanship. The project must also be feasible in terms of time, labour, and material to enable the builder to comply with contract requirements.

8.3 Materials for external walls

The following list of factors affecting the choice of materials for external walls will be given here without comment, but will be dealt with further as the need arises. The sequence is that used for the primary functions under the previous heading.

STRENGTH. The exterior wall should be capable of supporting own weight; strong in tension, compression, shear, or torsion as required; joints and jointing materials should match strength of wall units; stress should be confined within the capability of the material.

STABILITY. Movement due to applied loading, moisture content, temperature, thermal, and other causes should be structurally acceptable; plastic deformation should not be excessive. Walls should resist wind force; joints should be as strong as the surrounding material and absorb surface movement.

WEATHER EXCLUSION. Materials should be water-repellent, damp- and frost-resistant, and either totally or partially impermeable as required; jointing should be flexible enough to accommodate movement yet prevent capillarity; walls should be self-weathering, rendered, surface-coated, or with dry cladding added as specified.

THERMAL INSULATION. Walls should be designed against heat loss, solar heat gain, overheating or interstitial condensation; thermal transmittance or conductivity of materials should be calculated to meet specific requirements; ventilation and air circulation should be allowed for.

SOUND INSULATION. Basic principles should be observed concerning acoustic properties, porosity, sound absorption, noise control, including traffic, vibration, and effect of openings in walls.

DURABILITY. Weathering properties of materials and effect of atmosphere and climate should be considered; precautions against deterioration, corrosion, decay, chemical action, damp and erosion, lichens, mould, and fungal growth should be taken. Protective coatings and preservative treatment should be evaluated.

FIRE-RESISTANCE. Regulations and by-laws should be observed; fire grading of materials, openings in walls, fire doors, access, circulation, ventilation precautions, alarms, etc. should conform with requirements.

APPEARANCE. The effect of design on environment should be envisaged; changes due to atmospheric exposure, weathering, staining of brickwork, efflorescence, algae, and lichen growth, etc. should be anticipated. Changes in appearance due to cleaning, renovations, maintenance, applied surface-finishes, pointing, painting, rendering, etc. should be considered.

FEASIBILITY. Selection of appropriate factors to be made as listed above; assessment of priorities and compromise to be listed.

8.4 Solid and cavity wall details

The traditional method of constructing solid walls has been to use either stone, brick, blockwork of concrete or similar construction, usually under the blanket term of 'masonry', though a mason is strictly a builder in stone. The units are usually bedded in horizontal joints of mortar which permit the applied load to be evenly distributed throughout the wall. The mortar normally consists of sand with a binding matrix of lime, Portland cement, or both. Vertical joints usually act only as a filling.

The use of waterproof additives to mortar is a comparatively recent innovation. Without admixtures, joints between adjacent blocks tend to shrink or crack after drying out; this causes the inside of the wall to become damp and because of this, the use of solid walling is now less prevalent, particularly in brickwork. In housing and medium-scale construction where brickwork is most used, it is customary to form an air gap or cavity between the inner and outer skins, at the same time taking precautions to prevent the passage of water between the two. To ensure stability between them, it is usual to place wall

82 | Construction Technology

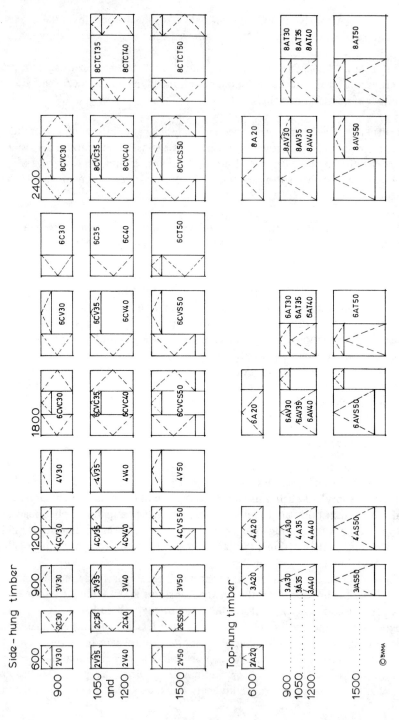

Fig. 8.4 Side- and top-hung timber casement windows from the E.J.M.A. Certification Trademark range

Functions of the 'external envelope' | 83

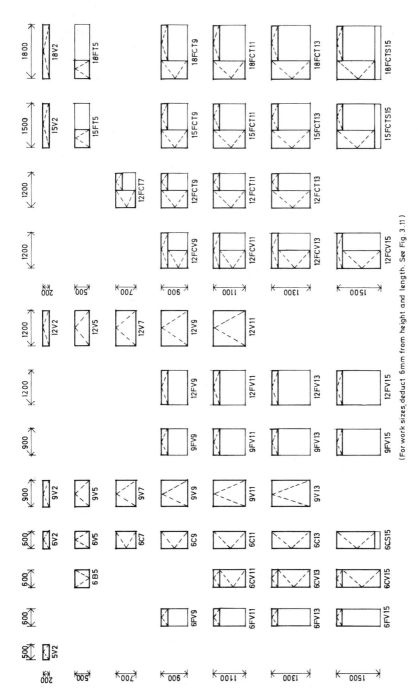

Fig. 8.5. Side-, top- and bottom-hung steel casement windows from the S.W.A. range. (B.S. 990: Part 2.)

(For work sizes, deduct 5mm from height and length. See Fig. 3.11)

ties across the gap at regular intervals; these can be of various shapes, but the most popular is the 'butterfly' (Fig. 8.2(d)). This has been designed with the object of preventing the passage of water and to provide as small an area as possible on which mortar droppings can lodge.

There is a wide variety of external types of walling, some of which are shown in Fig. 8.1. The most popular bonds for solid walls are English and Flemish as shown in Fig. 8.2. An example of corner bonding for concrete blocks is also shown. All the above are currently in use mainly as boundary walls, with the exception of Fig. 8.2(d). Further details will be given at Level 2.

8.5 Performance requirements of windows

The primary function of a window is to let in light but other considerations also arise including the transmission of heat and ventilation, i.e. satisfactory air circulation.

DAYLIGHT. The control and measurement of daylight has received close attention from the Building Research Establishment; its application to building is essential to the study of the architectural and building professions. Briefly, the amount of indoor light needed for a specific purpose can now be determined and is normally expressed as a ratio of indoor to outdoor illumination commonly known as the *daylight factor*. This depends on sky conditions, the size, shape, and position of windows, and the reflectivity of indoor surfaces.

SOLAR RADIATION. A further phenomenon of sunlight is solar radiation, both ultraviolet and heat-producing. Solar radiation raises the surface-temperature of the construction above that of the surrounding air. Thus, when the effects of solar radiation are considered in design calculations a compromise has to be made to take account of air temperature also.

Transmission of light alone through glass can be fairly easily controlled; but the material itself raises heat problems. Thermal transmittance through glass can be regulated, usually by use of single, double, or triple glazing (Fig. 8.3). Protection can be had from a sealed air space width of only a few millimetres, though for sound-insulation a wider space is needed. Tinted or sealed units are effective against both light and heat; and solar-control glass is also available to reduce the transmission of solar heat.

AIR MOVEMENT. In the warmer areas of Britain some form of cross-ventilation may be necessary; hence some buildings in holiday resorts, especially sun lounges and patios, are made as open as possible. The introduction of the glass louvre has been perhaps the greatest single advance in window design for such a purpose. In other areas, however, ventilation needs to be restricted to counter the effect of high wind; in northern climates it is curtailed by the window size necessary

Functions of the 'external envelope' | 85

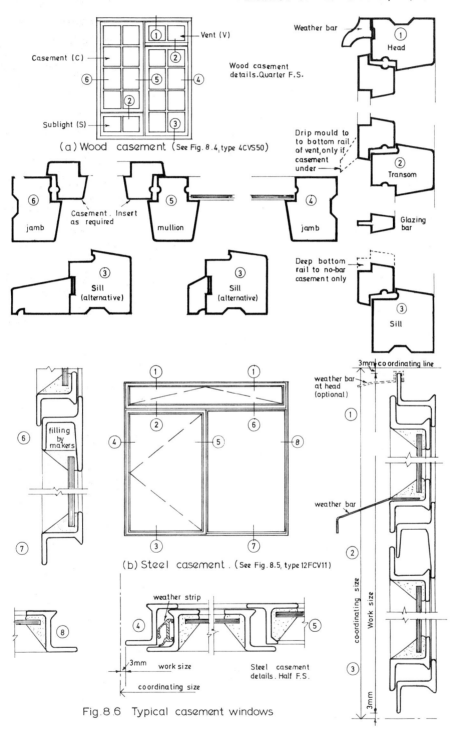

Fig. 8.6 Typical casement windows

to exclude colder rain or dust. Frequently, simple ventilation is insufficient to cope with human needs, for no natural method can make a building cooler than the lowest shade temperature.

In busy towns air movement can also be restricted where windows must be kept closed because of traffic noise; in such cases some form of mechanical cooling is installed, usually fans, plenum systems, or air-conditioners, though the latter can be noisy if of the unit type, or if not properly insulated.

NOISE- AND SOUND-INSULATION. This has been mentioned under *sound* insulation but is more applicable to windows than other parts of the structure. The use of double glazing for both sound and thermal control can be effective but application requires some study. B.R.E. 140, *Double glazing and double windows* gives much relevant information (see Fig. 8.3).

WINDOW DESIGN. In recent years this has become a task for specialists, as there is now a wide range of materials available to ensure that the windows are of a satisfactory standard, and that adequate performance requirements are met. It is now generally stipulated that the design should meet the appropriate British Standard or Code, especially those dealing with air infiltration (ventilation and draught) for varying conditions of exposure. The glass must be of appropriate thickness and resistance to water penetration is also defined to ensure that gross leakage does not occur. These considerations will also be covered at another level.

8.6 Conventional types of window

A traditional wood casement of straightforward design which has been in use for many years is shown in Fig. 8.6(a) which conforms to B.S. 644: Part 1, 1951. This specification covers the quality, design, and construction of casements to open outward, both with and without glazing bars. Similarly, double-hung sash windows, English and Scottish, are covered by B.S. 644: Part 2, 1958 and Part 3, 1951 respectively, and will be treated at Level 2.

A selection of sizes of side- and top-hung timber casements from the E.J.M.A. Certification Trademark range is given in Fig. 8.4. The identification marks and elevation, viewed from the outside, give all the necessary information once the code is understood. The abbreviations used are as follows:

V   top-hung vents over fixed lights
C   side-hung casements opening outward
A   top-hung casements
S   bottom sublights
T   top-hung vents over top-hung casements
    (Fixed lights have no prefix. The apex of the broken line triangle points towards the hinge.)

Using 4CVS50 identification in the 1200 mm range (Fig. 8.4):

4   width when multiplied by 300 mm module. i.e. 1200 mm
C   side-hung casement opening outward
V   top-hung vent over fixed light
S   bottom sublight
50 height of 5.0 when multiplied by 300 mm module. i.e. 1500 mm

STEEL CASEMENTS. A typical selection of steel casements taken from the S.W.A. range (B.S. 990: Part 2, metric units), is shown in Fig. 8.5. The identification marks for these frames are:

F   fixed light
C   side-hung casement opening outward
V   top-hung casement opening outward, full width of unit
T   top-hung casement opening outward, less than full width
B   bottom-hung casement opening inward
S   fixed sublight.

Using 12FCV11 identification in the 1200 mm range (Fig. 8.5):

12 width when multiplied by 100 mm module. i.e. 1200 mm
F   fixed light
C   side-hung casement opening outward
V   top-hung casement opening outward, full width of unit
11 height when multiplied by 100 mm module, i.e. 1100 mm

Window identification sizes are based on multiples of 100 mm, the vertical increment being generally 200 mm and the horizontal increment 300 mm; pressed steel filler panels of 100 mm increments make up the intermediate sizes. Casements in metric units, whether wood or steel, are deemed to be *handed* either left or right, according to the side on which they are hinged when looking from the outside.

8.7 Typical casement-window assemblies

Modern windows are made in a variety of types and materials. They may be side-hung, top- or bottom-hung, louvred, horizontal or vertical sliding, or combinations of the above. They may be coupled together by means of panels, transoms, or mullions to create larger glass frontages in a variety of materials. They are made of timber, steel, aluminium, plastics, glass fibre, or a combination of these. Double glazing can be inserted in most types without difficulty if required. Only two types will be described here.

A typical assembly of a conventional wood casement window with its component parts, according to B.S. 644: Part 1, is shown in Fig. 8.6(a). Standardiza-

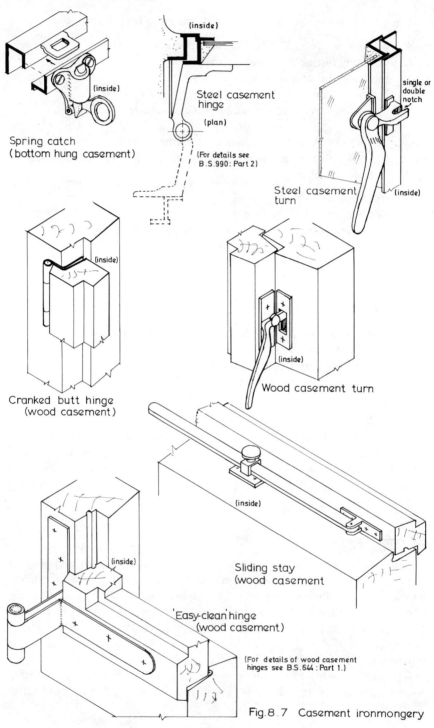

Fig. 8.7 Casement ironmongery

tion was based on the work of the English Joinery Manufacturers' Association, now the British Woodworking Federation. The details are shown a quarter full size, but the actual dimensions of the sections can be obtained from B.S. 644: Part 1. It will be seen that where opening casements are not required, fixed lights are formed by glazing direct into the frame.

A typical steel casement assembly is shown in Fig. 8.6(b). The names of the component parts correspond to those of wood casements. The details are shown at half size. It will be noted that the work sizes are 6 mm less than the co-ordinating sizes, as shown in Fig. 3.11.

IRONMONGERY. The more usual items of casement ironmongery for conventional windows are shown in Fig. 8.7. The most important of these is the hinge, which may be identified as any piece of equipment which enables one part of a component to pivot on another. It is essential when designing to ensure that the right type of hinge is selected for its task. The two types shown for wood casements have been sketched from plans given in B.S. 644: Part 1. Projecting hinges are used for both wood and steel casements to permit easy cleaning from the inside. They are usually of steel; and with metal casements they are normally supplied as part of the frame.

Typical types of casement fastener for conventional windows of wood and steel are also shown. For timber it is normal to use a hook plate with a flush frame, or a mortice plate with a rebated one as illustrated. A typical metal casement turn is also shown; this may have a single or double notch to enable ventilation to be controlled without impairing security. A typical spring catch for a bottom-hung steel casement is also given.

Casement stays are made in a wide range. The cheapest and most common is that of a peg stay but is not very secure; the sliding stay as shown is much more satisfactory. The screw-down knob tightens on the stay to hold it secure in any position. The knob may also be spring-loaded to engage in holes in the stay and is popular with both wood and metal windows.

9 | Doors

9.1 Typical performance requirements

The door is the usual means of approach or access to a building, room, or passage. It is the most constantly used moving component in the whole structure, particularly when placed at a busy access, and is often subject to constant use throughout its life. It can range in quality and construction from a makeshift barrier to an imposing entrance.

Doors can be made of wood, metal, plastic, flexible rubber, or any combination of the above. They can be side-hung, sliding, folding, shuttered, pivoted, or revolving and can range in size from small cupboards to the huge sliding folding-doors of aircraft hangars.

Standard types of doors and frames, now in almost universal use, are manufactured by specialist firms, whose designs are usually the outcome of long experience; it is advisable, therefore, to follow the makers' assembly and fixing instructions as far as possible.

Performance requirements vary greatly according to the location and degree of use. The primary function of doors may be conveniently listed under the basic headings already referred to for windows:
    strength and stability,
    weather exclusion,
    thermal and sound insulation,
    durability,
    appearance,
    feasibility.

STRENGTH AND STABILITY. It is often the case that the more functional the door, the less refined is its construction, and in the interest of prestige, the more imposing the entrance, the less it appears to be used. Service doors to stores and yards in industrial premises, for instance, are often subject to rough treatment in addition to the stress imposed by the weather. The stability of a door is related to the material used in its manufacture and the type of finish required; its strength will usually depend on the method of assembly, the framing of its members, and the protection it receives in transit, stacking, and storage, as well as the furniture and fittings applied after completion.

WEATHER PROTECTION. The function of a door is often related directly to its cost, but to a large extent its stability depends on exclusion of the weather. In exposed positions, precautions are taken in the form of grooves and throating. Weather strips, bars and draught-proofing, where provided, must be flexible enough to counter expansion and warping to allow the door to be opened and closed freely.

THERMAL AND SOUND INSULATION. Door insulation is usually a matter of dealing with sound rather than heat, particularly where privacy is important or noise has to be controlled. For good sound insulation the door must be solid and tight-fitting with draught-proof seals around the edge. For a high degree of sound control two doors enclosing a vestibule is the best answer if this is possible.

DURABILITY. Apart from good construction, proper maintenance of doors is needed, depending on the location and the degree of exposure to which it will be subjected. As external doors opening inward are normally placed on the inside of the jamb, they receive some protection from the weather, but external timber doors opening outward are especially vulnerable at the top and bottom edges, which, being out of sight, often receive the least maintenance.

FIRE-RESISTANCE. Fire regulations can be quite strict with respect to door sets, particularly in regard to structure, means of escape from fire, and prevention of smoke spread. Fire-resistance is usually assessed by estimating the time it takes for the flame to destroy the door's stability (collapse) and integrity (passage of flame). These will be covered more fully together with building regulations at another level.

APPEARANCE. As the entrance is usually the focal point of the structure, every effort is made to make it as attractive as possible. The design of an entrance door or its selection from a manufacturer's list is not always a straightforward matter, and its appearance usually has to be integrated with its function. The former will depend on the materials of which the door is made and whether it is self-finished or painted. Most doors in general use are painted, and a wide range of good quality finishes is available.

FEASIBILITY. As with windows, the feasibility of the type of door selected must be assessed by the fitness of the design for its intended purpose, i.e. the performance requirements outlined above. The final result will usually be a compromise of appearance, economy, and function. The last of these is normally the governing factor in the final choice.

9.2 Basic door types

Performance demands of doors tend to divide them into two broad types — those for pedestrian use and those for industrial use. The latter category may

92 | *Construction Technology*

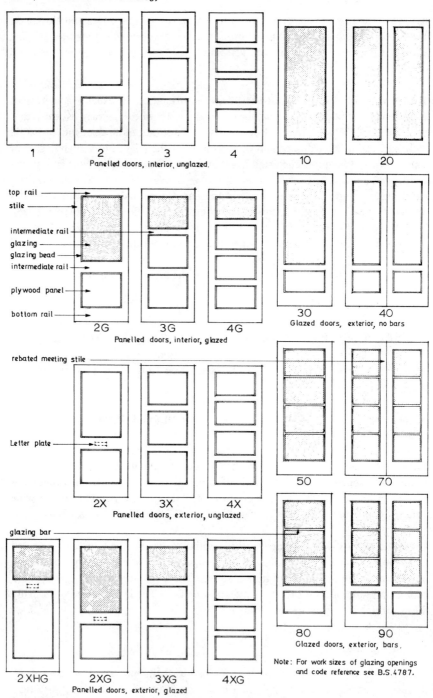

Fig. 9.1 Panelled and glazed wood doors. B.S 459. and 4787: Part 1

Doors | 93

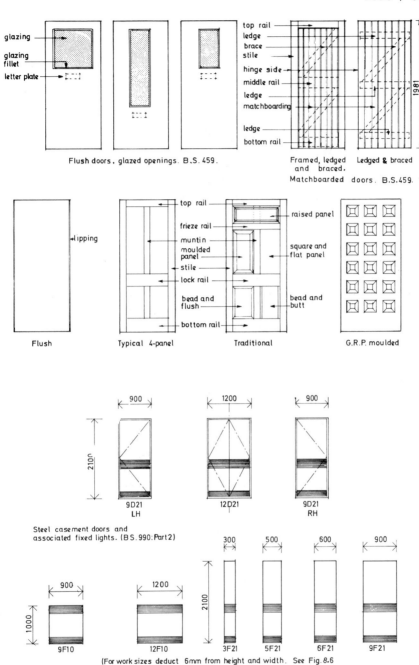

Fig. 9.2 Door types contd.

94 | Construction Technology

be assumed to include all larger doors permitting movement of vehicles, and these frequently employ special methods of operation such as collapsible gates, rolling shutters, flexible and garage doors, and similar. Only pedestrian doors will be covered here; these are of three basic types, unframed, framed, and flush doors.

UNFRAMED DOORS. These are the cheapest of all doors to make and are of the simplest construction. They are mainly used for temporary work or where security and appearance are not of high priority. The most basic of these consists of a sheet of timber-based board to which horizontal ledges are nailed to stiffen it and to provide a base for latch and hinges.

Ledged and battened door. This consists of tongued-and-grooved matchboarding strengthened by horizontal ledges but without braces. It has a tendency to warp diagonally and also to drop on the latch side owing to lack of bracing.

Ledged, braced, (and battened) door. This is more satisfactory than the one above, as the brace tends to prevent the distortion of the door (Fig. 9.2). This is possible, however, only when the brace is pointing *downward* towards the hinge. Should the door be inadvertently hung upside-down, the braces will point the other way and the door will tend to sag on its hinges.

Framed, ledged, and braced door. This door differs from the ledged type in so far as the stiles and top rail are made thicker than the battens and are strengthened at the intersections with adjoining members by means of properly constructed joints. Such doors are of a traditional type and have been in use for many years (Figs. 9.2 and 9.5).

FRAMED DOORS. Although the last named is strictly a framed door, it is usual to reserve this description for the traditional panelled types as shown in Fig. 9.1. The panels may be either solid or glazed, and those shown are in accordance with B.S. 459 and 4787: Part 1. A wide range of proprietary doors is now produced by manufacturers, and these will be covered at another level. Typical four-, five-, or six-panel doors are still built in the traditional style but are expensive compared with British Standard designs. Modern doors of moulded GRP are now being adopted for entrances to houses and medium-scale building (Fig. 9.2).

STEEL-CASEMENT DOORS. A range of steel-casement doors and associated fixed lights is shown in Fig. 9.2, in accordance with B.S. 990: Part 2. They follow the same design and construction as those shown for casement windows (Chapter 8).

FLUSH DOORS. In order to afford freedom in development and construction, the B.S. specification on flush doors, B.S. 459: Part 2, does not require any particular form of construction but specifies dimensional design only. It contains

Doors | 95

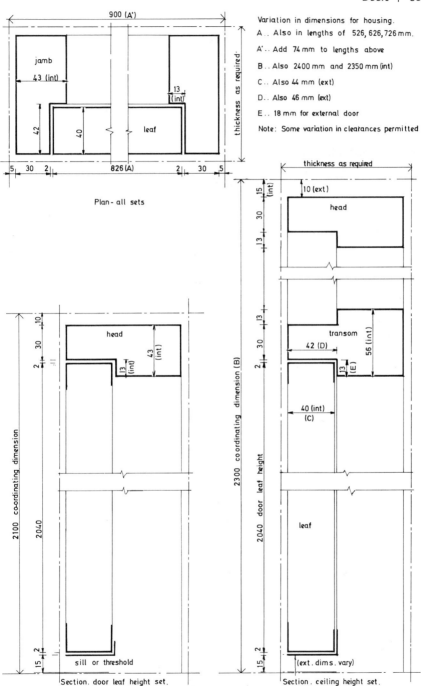

Fig. 9.3 Standard dimensions wood doorsets   B.S 4787: Part 1

no provision for strength and stability as this is left open to agreement between the supplier and purchaser. Details of glazed openings (Fig. 9.2), will be given at Level 2.

9.3 Sizes of doors, door sets, and openings

The subject of component design and tolerance is covered by the appropriate British Standards and will not be discussed at this stage. It must be mentioned, however, that when the designer submits the basic size of the components to the manufacturers, the latter must allow for gaps and tolerances to ensure that the components can be readily manoeuvred into position.

SIZES OF DOORS. The standard sizes of doors given below, based on B.S. 459, have been converted directly from imperial measurements (Figs. 9.1 and 9.2). These are not to be confused with metric door dimensions given in Fig. 9.3.

Table 9.1 *B.S. 459: Part 1: 1954. Panelled and glazed wood doors*

Type	Height	Length	Thickness
Interior	1981	610 686 762 838	35
Exterior	1981	762 838	44
Glazed	1981	762 838 914	44
Garage	1981	2134	44

DOOR SETS AND OPENINGS. Figure 9.3 shows the dimensions of a standard metric door set designed to fit into a basic space of 900 mm wide and basic height of 2300 mm for ceiling height set, and 2100 mm for door-leaf height set. These are typical dimensions but the full range of sizes according to B.S. 4787: Part 1, is as follows:

Basic ceiling heights	2300, 2400, and 2350 mm
Door-leaf height	2040 mm
Door lengths	526, 626, 726, and 826 mm
Door thickness	40 mm internal, 44 mm external
Rebate in frame and fanlight	13 mm internal, 18 mm external
Gap between head and ceiling	15 mm internal, 10 mm external
Gap under door sill	15 mm internal, external varies

Using the above alternatives and taking any door size as a base, the following dimensions would obtain using a door leaf height set:

Doors | 97

Door opening height	+ 4 mm (± 1 mm)
Frame height	+ 50 mm
Structural height, basic	+ 60 mm
Door opening width	+ 4 mm
Frame width	+ 64 mm
Structural width, basic	+ 74 mm

Table 9.2 *B.S. 459: Part 2: 1962. Flush doors*

Type	Height	Length	Thickness
Interior	1981	610 686 762 838	35
Exterior	1981	762 838	44

For *B.S. 459: Part 3: Fire-check doors* see Level 3

Table 9.3 *B.S. 459: Part 4: 1965. Matchboarded doors*

Type	Height	Length	Thickness
Ledged and braced	1829 1981 1981 1981	610 610 680 762	28
Framed, ledged, and braced	1981	686 762 838	44

9.4 Typical door-frame assemblies

Typical matchboarded doors in accordance with B.S. 459: Part 4, are shown in Fig. 9.2; shown also in plan in Fig. 9.5, where a non-standard frame has been added. This is normally of wrot timber with stops nailed to the jambs only.

DOOR-FRAME ASSEMBLIES. Typical details of external doors and frames are given in Fig. 9.4. The frame may be held in position either by iron dowels let into the step, or by the threshold sill as shown. Frames may be either built into the wall as the work proceeds by means of galvanized brackets or cramps screwed to the back of the frame. In order to protect the frames during transit, horns or haunches are often left projecting, and these may be either sawn off when fixing or left in position if preferred.

With unframed doors the rebate stop is usually planted on as shown in Fig. 9.5 or it may be worked on the solid as in Fig. 9.4. Planting is more economical; it also permits the stop to be adjusted to suit the door after it has been hung

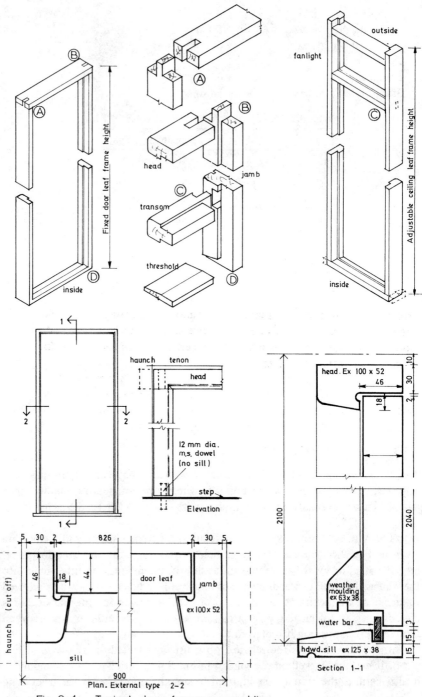

Fig. 9.4 Typical door frame assemblies

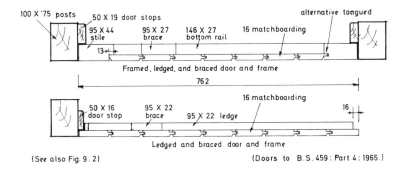

Fig. 9.5. Door frame assembly. Matchboarded doors.

but is not so strong. A typical section through an external door and frame is shown in Fig. 9.4, which also includes a weather moulding as specified in B.S. 459: Part 1. This can be used for both a rebated or unrebated bottom rail.

9.5 Ironmongery

Ironmongery to doors is usually classified either as fittings which allow movement, such as hinges, or those which give security, such as locks, latches, and bolts. In addition the door may be supplied with *furniture* such as kicking- or finger-plates, letter plates, etc.

HINGES. The most common type of hinge is the *butt* (Figs. 9.6 and 9.8). Normally one pair measuring 75 or 100 mm is sufficient for internal doors, though heavier external doors usually have one-and-a half pairs, i.e. three hinges. They are usually made of steel, though brass with steel pins is also used. In order to avoid cutting a recess for the butt, as shown in Fig. 9.6, thinner hinges of bright steel are now in wide use. These permit surface fixing, which is much more satisfactory (Fig. 9.8). In order to lift the door clear of the mat or carpet as it opens, *rising butts* are sometimes used, which also act to some extent as a self-closing aid (Fig. 9.6).

For ledged or unframed doors, *tee* or *cross-garnet* hinges are in wide use, especially where the door may be subjected to rough treatment, (Fig. 9.6). Strap hinges serve a similar purpose. They may be of single or double leaves, the latter being much stronger and often used for framed garden gates. The supporting pin is usually secured to the post or pier, which allows the gate to be lifted off its hinges.

LOCKS AND LATCHES. There are four basic types of locks and latches.

The dead lock, which has no handle and is operated by a key.

The mortice latch, which has no lock and is operated only by a handle which moves a spring-loaded bolt.

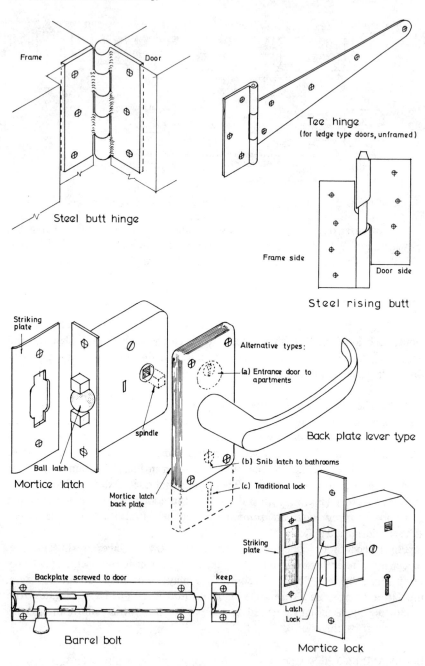

Fig. 9.6 Door ironmongery

The mortice lock which normally contains both dead lock and mortice latch and is commonly called a *lock-set*.

The night latch or cylinder rim latch. This was once almost exclusively referred to as a *Yale* and is now in very wide use. Cylinder latches are compact and require only a small key. They are usually fitted with a thumb slide on the inside which can hold the bolt open or lock it shut if necessary to prevent the key from being used.

Back-plate lever. This combined plate and handle, now in wide use, is shown in Fig. 9.6. The back-plate is obtainable in a variety of styles; one type designed for entrance doors to apartments takes a Yale-type key; another has a snib latch for bathrooms, etc; another has a key-hole for a traditional mortice lock. Most back-plates to entrances have a spring-loaded snib on the inside to hold back the lock when it is not required. It is also used with mortice latches as shown, particularly for internal doors to dwelling houses where locks are not required.

Door handles can also be of the knob type, though levers are now much more

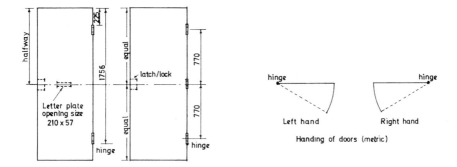

Fig. 9.7. Position of locks and hinges

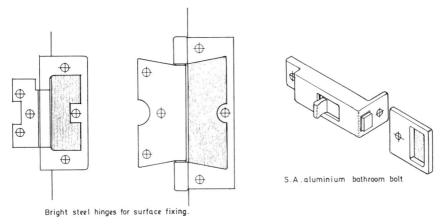

Fig. 9.8. Door ironmongery

popular. The handle itself engages the spindle which turns the latch. Usually the back-plate is screwed to the face of the door and the length of the spindle varied according to the thickness of the door.

Bolts are available in a variety of styles and materials. The normal type is the *barrel bolt* (Fig. 9.6) usually placed at both the top and bottom of the door. A smaller type of S.A. aluminium bolt is also popular for bathrooms, w.c.s, etc. (Fig. 9.8).

POSITION OF LOCKS AND HINGES. The standard position of hinges for both flush and panelled doors is shown in Fig. 9.7 together with that of the latch or lock. The standard position of the opening for letter plate and postal knocker for unglazed flush doors is also given.

Handing of doors. In addition to the correct position of locks and hinges as shown above, it is necessary to know the *hand* of the door, i.e. its left or right side in relation to the direction in which it opens. The convention for a metric door and window (not imperial) is that it is left- or right-handed according to the hinge position when *pulled* open (Fig. 9.7). The imperial convention is the reverse of metric. It will be seen from Fig. 9.6 that the mortice latch can be used for either hand but the hand for the mortice lock must be stated.

10 | Basic roof forms

10.1 Performance requirements of roofs

The concept of the roof structure in relation to the external 'envelope' has been discussed in Chapter 8. When built as a dome or tent it can become the whole 'envelope', but more usually it is supported by walls or columns. It can be either flat, pitched, or curved in shape and the forms of roof now in common use are of great variety. Each is designed to perform a specified function and, as with other elements of building, the components of the roof will have to fulfill a number of requirements to meet a minimum standard of performance. In order to do this a *performance specification* in the form of a check-list has been produced by H.M.S.O. termed *Performance specification writing for building components, D.C.9,* which is a summary of the main requirements. This goes into some detail, but at this stage it is sufficient to follow the pattern of discussion employed in previous chapters under the following headings:
 strength and stability,
 weather exclusion,
 thermal insulation,
 sound insulation,
 durability,
 fire resistance,
 appearance,
 feasibility.

STRENGTH AND STABILITY. The composition, manufacture, and assembly of the basic roof-structure are primary factors affecting strength and stability as much as the form of the roof itself. These will be influenced by the shape, dimensions, geometric properties, and weight in addition to structural properties, such as resistance to tension, compression, bending, and fatigue. It may also have to resist mechanical wear, particularly if used as a support for moving loads.

Shapes affecting strength. A great deal of what has been considered in Chapter 7 could also be applied to roofs, especially where *stiffness* is an important factor. Such shapes can also be affected by size or span and by the loads they have to carry. Roofs are generally referred to as *short-*, *medium-*, or *long-*span: short-span roofs are those up to 7.60 m, medium-span are between 7.60 and 24.40 m, and long-span are over 24.40 m.

In all types of structure it is necessary to keep the dead weight to a minimum so that imposed loads may be carried with greater economy of material. Short span construction is obviously the cheapest, and in units of small cells such as those shown in cross-wall construction (Fig. 7.5(e)), where intermediate supports form part of the building, the solution is fairly simple. But with large modern structures, such as stadiums, assembly halls, depots, hangars, etc., where columns cannot be used, the problem of weight becomes significant.

Wind- and air-pressure. As shown in Fig. 7.2, wind blowing square against a building is slowed up, with a consequent build-up of pressure against the face; at the same time it is deflected and accelerated around the end walls and over the roof. As this can cause damage to glazing and cladding, local authorities are increasingly insisting on adequate calculations deemed to satisfy the wind-resistance requirements of the region.

Roofs with a pitch of 30 degrees or less may be subjected to severe suction, but above this, positive pressure can be built up, though even here suction can develop at ridge level. Suction can be in excess of the dead weight of the roof, which would then require firm anchorage to an adequate foundation. Distribution of wind over a roof is not uniform; wind may blow along a ridge, or vortices may be produced along the edges of eaves and verges when the wind blows against a corner. High suction can thus occur at all edges and corners of square buildings.

Wind loads. Details of calculating wind loads are given in a number of documents, the best known of which is the British Code of Practice, *3:* Chapter V: Part 2. This states the various factors which should be taken into account when designing structures and components. When wind blows over roofs of low buildings the area is sheltered: when it meets taller buildings, however, a different flow-pattern occurs. Below, it descends to form a vortex in front of tall buildings and to sweep around corners. If the building is on stilts or has an open ground plan, the descending wind passes beneath. If the structure is not too high, it can be refreshing, provided it is not cold and is free from dust. Unpleasantly high wind speeds in the environment of the building must be avoided, but where buildings have already been erected, steps may have to be taken to alleviate the worst effects.

WEATHER EXCLUSION. Protection from the weather is chiefly concerned with rain, though wind, snow, and sunshine must also be considered. The entry of water, however, is the main problem. The roof structure and materials are usually related to properties of moisture content, capillarity, absorption, penetration, drying and evaporation, and movement due to heat and moisture. Resistance to water penetration, however, is the chief consideration, and roofs are usually designed with this property in mind. Roof coverings and finishings together with drainage will be covered in Level 2.

THERMAL INSULATION. Although this subject is concerned with thermal transmission through the external 'envelope' as a whole, the problem lies mainly with the roof. A study of thermal properties includes movement, freezing, melting, radiation, conductance, and temperature, all of which affect the roof structure. Thermal insulation, however, is rarely involved in the choice of roof types, though finishes can affect it. Usually the solution lies in the nature of the insulation material itself, which may be a flexible layer laid on or under the roof, or a self-supporting slab strong enough to act as a sub-structure. Lightweight concrete is also used where the structural design permits.

SOUND-INSULATION. As mentioned briefly in Chapter 8, roofs need to be designed according to their purpose, location, and construction. The principal factors to be considered are the sound-insulation properties of the material used and the completeness and continuity of the 'envelope'. (Air gaps in pitched roofs, for example, reduce the potential insulation.) Materials used to improve thermal insulation invariably improve sound-insulation also. Figure 10.1, reproduced with permission from B.R.E. *Digest* 128, shows the effect of outdoor noise on buildings in areas vulnerable to aircraft traffic, and how materials and construction assist in sound-reduction. Hospitals and hotels usually have special requirements and lecture-, concert-, and conference-halls often need acoustic treatment as well as absorbent material.

DURABILITY. The durability of a roof will depend partly on its structural properties and stability, but chiefly on its finishes and the maintenance it receives. Roof drainage and freedom from blockage play a vital role in the satisfactory performance of the covering. Resistance to wind and driving rain must also be considered. Some materials have a limited life and are used initially in the interests of economy; others can be affected by industrial atmosphere or conditions in coastal areas. Weather exclusion and durability are closely related.

FIRE-RESISTANCE. The effect of fire on roofs can often be related directly to the degree of combustibility of the material, fire-resistance of the structure as a whole, surface spread of flame, effect of other building materials, or change in behaviour or movement during use. The form of construction should also be designed to prevent the passage of fire through the roof structure itself. The hazard of fire spreading to roof from a nearby fire is dealt with in B.S. 476. Fire-resistance will be discussed together with building regulations at another level.

APPEARANCE. This will depend on the basic type of roof form, the materials used, and the purpose for which it is intended. Sometimes, however, good functional design is sacrificed in favour of a more acceptable shape or finish or, when strength, stability, durability and economy are deemed less important than appearance. Tradition in design can also affect appearance, and this is

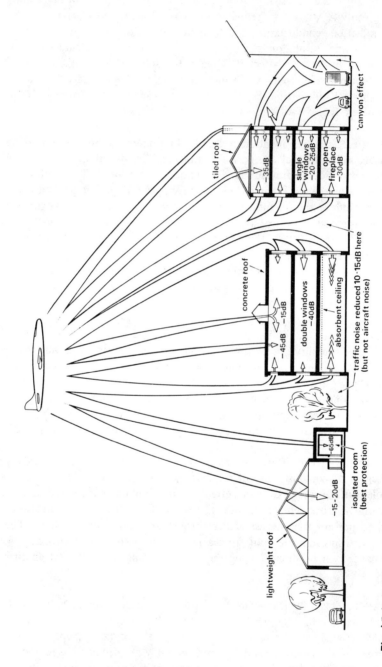

Fig. 10.1 Buildings exposed to outdoor noise

clearly a matter for the designer and client rather than for the builder. This does not apply to prefabricated elements and components and industrial buildings delivered to site, of which there is now a range of choice.

FEASIBILITY. The feasibility of roof design and construction is often governed by factors which apply to the external 'envelope'. In addition, the use of the roof for purposes other than covering could also be a matter of feasibility as well as conformity to building regulations, which for roofs can often be quite demanding.

10.2 Basic roof forms

Basic types of structure, as discussed in Chapter 7, have a direct bearing on roof forms, particularly when walls and roofs form part of the same surface, as with domes and vaulting. The majority of roofs are either *pitched* or *flat* and a selection of roof forms is given in Fig. 10.3. Many other roof types are in general use, some of which are referred to by the shape of their surface, such as barrels, while others are of shapes depending on structural principles designed to resist the forces applied to them, e.g. triangulated, framed, or trussed. They may also be referred to in terms of span. Only flat and pitched roofs will be considered here.

FLAT ROOFS. These may be of either timber, metal, concrete, or composite, all of which are now in wide use in multi-storey, commercial, industrial building and some dwelling houses. A single flat roof is simply one which spans between supports as shown in Figs. 7.1(a) and 10.3. This is normally built by placing timbers across the span at regular intervals, usually 400 or 600 mm. Should the width of the building be too great, intermediate supports can be provided to convert it into a double or triple roof (Figs. 7.1(b) and 10.2).

Roofs up to 10 degrees gradient are usually defined as flat. Where a slight

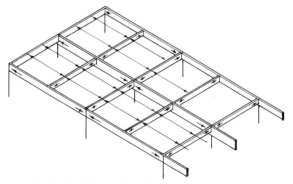

Fig.10.2 Stress distribution – triple roof

gradient is needed, say, 1:100, to carry off rain-water, thin timber fillets or firrings are used as packing to create the necessary fall. For concrete roofs screeds are usually used and these will be considered at Level 2.

PITCHED ROOFS. These are basically of two kinds: *single pitch* (or monopitch) and *ridge*.

Monopitch. This is a roof of single pitch, i.e. of single rafters inclined at an angle of 10 degrees or more, now much in demand in modern building, both industrial and domestic. It is economical in material, simple to construct, and the design can be adapted to suit varying needs such as car-ports, ridge beams, broken pitch roofs to provide loading bays, canopies, outhouses, etc. A common variant is the *lean-to* which, as the name implies, is a roof covering built against a higher structure. One advantage of the single pitch roof over the ridge roof is that roof loads carried by the walls are vertical (with some exceptions) and so reduce the risk of overturning or outward spread. Monopitch roofs can also be constructed by placing the rafters or purlins horizontally on supporting walls which are themselves built to a gradient.

Ridge roofs. a ridge roof is one where the roof slopes down from the apex, commonly called the ridge. This term is also applied to the horizontal member at the apex to which the sloping members or rafters are secured. It is possible to construct a pitched roof without a ridge piece by nailing the rafters together at the apex.

Couple roofs. These consist simply of pairs of pitched rafters nailed at the apex either with or without a ridge piece and secured at the base of the roof, i.e. *the eaves,* to prevent them spreading outward. This outward thrust of the roof is resisted by the wall, which needs to be strong enough to resist outward spread (Fig. 7.3(d)). This roof has the advantage of providing a clear internal roof space; the greater the pitch the more the headroom and the less the thrust on the outer walls.

Close couple roofs. In order to reduce the outward thrust of the rafters on the walls and to provide support for the ceiling, it is usual to insert horizontal timber ties or ceiling joists at eaves level which are nailed to the feet of the rafters. To prevent sagging they can be provided with intermediate support either from convenient internal cross walls or by the use of straps or timber hangers from the rafters inside the roof space.

Collar roofs. When cross-ties must be provided, but where extra headroom is also needed, it is usual to insert collars, not at eaves level, but higher up the rafters, and secured by nails or bolts (Fig. 10.3). This results in a partly sloping ceiling, often seen around the tops of walls in some buildings, especially bedrooms or buildings of a temporary nature. It is an economic and effective measure where sloping ceilings are not objected to.

Roof slopes. The greater the pitch or slope the less outward thrust at the eaves; also, a steep pitch sheds rain or snow more easily and the enclosed space

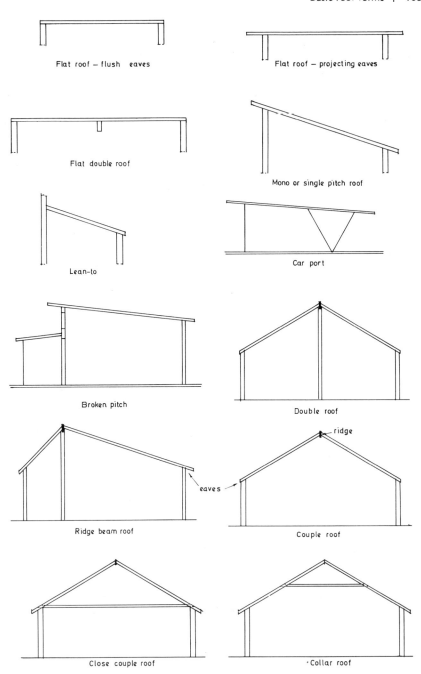

Fig.10.3 Flat and pitched roofs – basic types

can often be used as an attic or for storage. It is, however, more expensive to construct, and flatter roofs of 30 degrees pitch or less are now in general use.

The roof plan and its shape has considerable influence on its strength and one of the chief factors in effecting this is triangulation. Thus hipped ends are much stronger than gables but are more costly. Gables can be braced by nailing diagonal members on the underside of the rafters from the gable apex to the eaves (Fig. 7.2(c)). Pitched roof rafters can also be strengthened by trussing and it is now usual to use prefabricated trussed rafters of various sizes which are delivered to site and hoisted into position. These will be considered at Level 2.

D. Internal construction

11 | Elemental parts of internal construction

11.1 Walls, floors, and stairs as primary internal elements

When reference was made in Chapter 1 to the elements of the built environment, it was pointed out that the functional elements could be subdivided into component parts, such as internal and external walls or partitions, or joined-up units of floors, walls, or roofs to produce a structural frame: at the same time the primary purpose of building was to create a satisfactory internal environment. But the external 'envelope' in itself, though once considered sufficient for the provision of such an internal environment, is no longer enough to achieve this; internal space now has to be subdivided in order to contribute to the comfort and well-being of the living organism or to the efficiency of the machine. It is with these characteristics that internal walls, floors, and stairs are concerned.

The basic parts of internal construction are usually referred to as *internal walls* or partitions; *floors,* both ground and upper; and *internal stairs* leading from one level to another. The function of walls, floors, and stairs in relation to internal construction will be discussed in the following chapters. It is necessary here, only to identify those parts which refer to the primary internal elements.

The functional elements of internal walls, floors, and stairs are usually made up from smaller parts and from such materials as timber, stone, lime, gypsum, sand, clay, and ores, most of which are processed in some way from their natural state and converted into products suitable for building use. Timber is cut to lengths to produce planks, joists, boarding, etc. or processed into timber-based boarding of various patterns and thicknesses. Clays are tempered, moulded, burnt, and processed to produce building blocks, bricks and tiles, and floor and stair finishes. Stone for internal work can be sawn into slabs or crushed to provide aggregate, then mixed with sand and cement to produce concrete which is either cast *in situ,* or moulded into blocks, or made into pre-cast concrete products for walls, floors, and stairs.

Many other materials are used. *Steel* is converted into sheets, bars, and sections for floors and patent partitions; glass is inserted into partitions or used as a finishing; plaster is applied in the form of a film, or lightweight filling, or as wall coating, or mixed with glass fibre. *Component parts* of elements are used as *sections* in boarding, rolled-metal sheets; *units* as bricks, blocks, tiles, and panels; *compound units* as internal doors, windows, and stairs, while *compound unit components* are to be found as brick panels or timber-framed wall components and stairs. There are, of course, many more.

WALLS. An internal wall is an erection of brick, stone, timber, metal, plaster, plastic or any combination thereof, designed to enclose space. It may be either a *partition,* a *division* or *fire-break* wall, or *party wall,* each having its own function.

For internal walls:

A *partition* may be identified as a wall whose primary function is to divide space within a building or structure.

A *fire-break* is a non-combustible separating wall required to prevent lateral spread of flame. (In some areas this may be required to extend above adjacent roofs.)

A *party wall* is a wall common to two buildings.

Internal walls may be constructed *in situ* from units of brick, concrete block, or lightweight material, or produced in prefabricated component form in panels of storey height, in several vertical sections if necessary. Such components may be made either on-site or off-site. Although on-site fabrication is still used where circumstances permit, it is now common practice to use proprietary products, often fitted by the maker's experts with special jointing and fixing methods. Wall components may also contain built-in openings or divisions selected by the client from the manufacturers' catalogues. Factory-produced components of this type are competitive in price, of clean finish, and rapidly fixed. Assembly operations are usually carefully worked out to allow for tolerances, jointing methods, and dimensional control. An internal wall should not be confused with the inside or inner lining, skin, or shell of an external wall.

FLOORS. A floor may be defined as the lower supporting surface of a room or building. The expression is rarely used in external construction, other terms like veranda, balcony, area, platform, etc., being preferred. Like an internal wall, it can be an erection of clay block, concrete, timber, metal, or any combination of these.

The floor is usually referred to by its location such as basement, lower-ground, ground, second, or just as an upper floor. The height of a building is often referred to by the number of floors it contains. It should be noted that floor reference in Britain differs from that in USA and some other countries in that the first floor in the UK refers to one floor above ground level whereas in America the ground floor is called the first floor. In Britain this practice can lead to confusion if related to storeys, as in both Britain and the USA the first storey refers to the ground-floor rooms.

Where the ground is in contact with the surface slab and actually supports it in the manner described in Chapters 6 and 7, the floor is known as a solid floor; upper floors are referred to as *suspended floors.* Up to the time of the Second World War, it was customary to build suspended ground-floors of timber

Elemental parts of internal construction | 115

joists supported by dwarf walls, the object being to save infilling and prevent contact with the ground. Such floors, however, had to be ventilated on the underside, which usually led to underfloor draughts and high heat loss. Such floors are now less commonly used, having been superseded by the solid ground floor in which a continuous damp-proof membrane is usually inserted (Fig. 13.1).

Forms of construction. The construction of the floor will vary according to the structural concept on which it is designed, the materials of which it is built, and the mode of construction. Solid floors may be of concrete cast in one operation *in situ;* pre-cast concrete or hollow clay blocks can also be made in predetermined lengths and lowered on to cross-walls by crane. Pre-cast concrete beams or joists infilled with hollow pots or blocks are also widely used. Timber-joisted floors covered with floorboards are traditional, but floor panels are now superseding them. The form of construction will depend on a number of factors, chief of which are economy and function. These will be considered in Chapter 13.

STAIRS. To the functional element of walls and floors, the stairs themselves, which may be described as a number of steps connecting one level to another, must be added as a separate entity. Elemental forms of stairs once consisted of solid stone or timber baulks bedded in earth or spanning between walls. The practice evolved into the use of planks and slowly developed into the treads and risers which we know today. Stone steps were also cantilevered out from walls, a method much in use today, though concrete is now used (Fig. 11.1).

Stairs are usually classified according to the mode of construction. The main types are as follows.

Straight-flight stairs (Fig. 14.3). These have no landings and no turns.

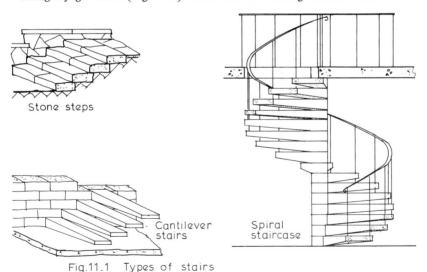

Fig.11.1 Types of stairs

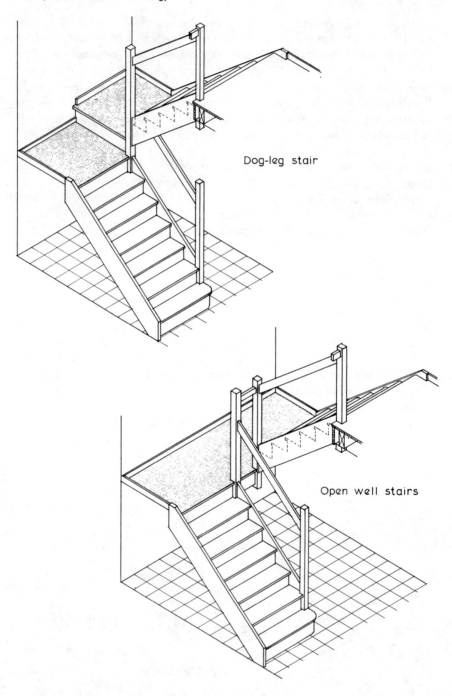

Fig.11.2 Types of stairs - timber.

Dog-leg stairs (Fig. 11.2). These consist of two flights which change direction usually half way up and which are connected by a landing or shaped steps. The object of these is to fit the staircase into the smallest space possible.

Open-well stairs (Fig. 11.2). These are similar to dog-leg stairs but have an open space or well between flights. They require more room on plan than dog-leg stairs, but give greater freedom at landings and permit a continuous handrail if desired.

Spiral stairs (Fig. 11.1). This is an age-old form of construction and is still to be seen in turrets and towers of medieval forts, castles, and cathedrals. It consists of a central newel as part of the narrow end of the tapered step. In recent years spiral stairs have revived in popularity. With precast concrete stairs the central newel is often made hollow to take concrete, the concrete is poured in and reinforced to form a solid vertical column. Spiral stairs are also made of timber, metal, or of composite construction often packaged and supplied k.d. complete, by proprietary firms.

More elaborate types of stairs are also in use, but only basic straight flights will be described in this volume. These will be found in Chapter 14.

12 | Internal walls: function and basic construction

12.1 Primary function of internal walls

The internal wall is often called upon to fulfil many of the functions of the external 'envelope'. Though not usually expected to be resistant to the weather, it is often needed to provide space enclosure, structural support, privacy, and some degree of comfort. It is also required sometimes to meet the special needs and functions of external walls, as described in Chapter 8. Internal walls must also be designed to meet whatever building regulations or standards are currently in force.

In Chapter 11 internal walls were classified under three main headings: *partitions; division* or *fire-breaks;* and *party walls.* Each of these may be subdivided into two types, *loadbearing* and *non-loadbearing.* For internal walls, four forms of construction are normally employed: masonry, monolithic, frame, and membrane.

A *masonry* wall is frequently of the loadbearing type, usually supporting a suspended floor, another partition above it, or both. It consists of units of clay or concrete brick or block, bonded together in a suitable mortar and may be used for all types of internal walling.

A *monolithic* or solid wall is almost always of concrete, either plain or reinforced, cast *in situ* or sometimes pre-cast and lowered into position by crane. This type is in use for party walls and in cross-wall construction.

Frame construction consisting normally of timber sections, are now in demand as partitions, party walls to dwellings, and also as inner leaves of external cavity walls. They are often covered with a dry lining or sheathing of plasterboard or timber. They may be either loadbearing or non-loadbearing (see Level 2).

Membranes are also used as internal partitions and are produced in a variety of ways. Usually they consist of two thin sheets of timber or plasterboard bonded on either side of a core of foamed plastic or other material and usually made to storey height.

The main functional requirements of internal walls are:
strength and stability,
thermal insulation,
sound insulation, and
fire resistance,
though not necessarily in that order. These requirements will also vary according to the group to which they apply.

Internal walls: function and basic construction | 119

STRENGTH AND STABILITY. The structural requirements of internal walls are different from those of the external 'envelope' where the nature of the stresses tend to vary. Overturning due to outward thrust or wind-resistance, for instance, would not usually be a serious possibility, whereas failure by buckling or lack of lateral support is not unknown. Failure could also occur by crushing, settlement, or movement of the foundation slab. Where internal cross-walls are used, damp patches may occur at intersections with external walls due to inadequate weatherproofing. Sliding and cracking could also take place due to horizontal pressure or settlement. Internal partitions could also fail around door openings as a result of shear forces and inadequate stability of the door-frame if subjected to heavy use.

THERMAL INSULATION. This is rarely a problem in internal construction as transfer of heat from one room to another is not usually of consequence, particularly where doors are in constant use, thus making control difficult. In storage areas, cold rooms, or laboratories where a constant temperature is desirable, special arrangements can be made to prevent heat loss or gain; these may also have to include provision against heat loss through ends of cross-walls, as these could be affected by damp.

SOUND-INSULATION. As mentioned in Chapters 8 and 10, the prevention of the passage of sound is often a significant factor and this may be as important in the internal wall as in the external 'envelope'. The effect of sound on human comfort is becoming more of a problem as towns expand and progress. This is due not only to traffic and mechanical and aircraft noise but also to artificial amplification of electronic equipment. Insulation against airborne and impact noise becomes increasingly necessary and considerable effort is being made to reduce disturbance to acceptable standards. Structural vibration in industrial building also needs special consideration.

FIRE-RESISTANCE. As will be seen from Chapter 8, the degree of fire-resistance in the case of walls, depends on a number of factors, including the need for walls to act as highly resistant fire barriers. They must also contain the fire within limits so that there are safe escape routes for the occupants. This is particularly true of internal walls between staircases and corridors which could be used as a means of escape. Fire-resistance of building elements will be dealt with more fully at another level.

12.2 Loadbearing and non-loadbearing walls

Non-loadbearing walls normally support only their own weight, which can be negligible, but loadbearing walls carry in addition dead and imposed loads from the floor and roof. With separating or fire-division walls, the dead weight can be

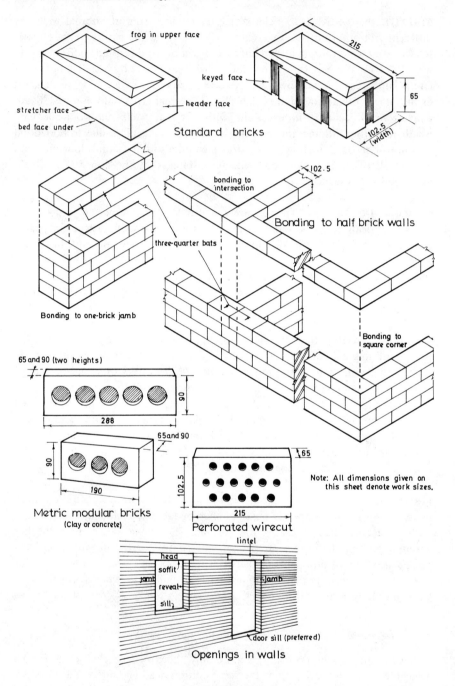

Fig. 12.1 Brickwork to internal walls

substantial. This also applies where noise requirements have to be met, though this can be done by other means, including the use of insulation materials. An internal wall can act both as an enclosing and supporting element, though this need not necessarily be so.

LOADBEARING WALLS. Where a wall is required to support a load, the method of dividing up the internal space must first be considered as this could affect the distribution of weight. The direction of the upper floor joists or floor slab for example, could decide the load bearing on the internal wall, and where two internal walls intersect at right angles, the stability of each could be increased. The plan shape of the structure is therefore important as is the storey height and wall arrangement to prevent buckling.

In the event of internal walls containing openings such as sliding partitions or folding-doors, different stresses could be set up, even though the partition itself may not be carrying weight from above. Here it may be necessary to consider means of supporting the top of the internal wall to resist buckling or overturning.

Where two internal walls or partitions meet at right angles, the method of bonding or securing one to the other is usually a simple matter (Fig. 12.2). (Stud partitions will be considered at Level 2.) Where the internal wall frame or membrane meets the external wall, it is sometimes not possible to tie in, particularly where some form of industrial system is being used. In such circumstances precast or preformed internal walls are normally made slightly shorter than the nominal width to allow for tolerance and are made sufficiently stable in themselves to prevent buckling. In some industrial systems the outer walls are supplied complete with inside finish with no provision made for internal partitions.

Foundations. Foundations for structural walls and slabs have been discussed in Chapter 6. Lightweight internal walls supporting small loads, as is usual in domestic building, require only thickening of the floor slab (Fig. 12.2), but in the case of cross-wall construction a normal foundation may be required, the size depending on the dead weight and imposed load on the wall.

NON-LOADBEARING WALLS. These are widely used as a method of dividing up space, and many proprietary systems are available, though the traditional method of using framed or stud partitions is still popular (see Level 2). This is covered usually with a dry lining of plaster or timber-based board, though occasionally it is plastered *in situ*. The framed partition is generally favoured in some forms of industrialized domestic building and is often loadbearing. For heavier construction lightweight hollow clay or concrete blocks are built *in situ* (Fig. 12.3) and can be made of various thicknesses to suit the room width and height. These can be made more stable by bonding in to the external wall (Fig. 12.2).

In industrialized building it is also usual to manufacture lightweight membranes to room width and height and lower them into position as the work proceeds. Provision is usually made in the external walls or upper floors to secure them. The most widely favoured form of internal non-loadbearing partition is usually one of the many proprietary products which will be dealt with at Level 3. It usually consists of panels of storey height and of width convenient for one or two men to handle and fit into grounds provided by the maker. It is neat in appearance, clean and speedy in erection, and can be quickly dismounted and re-erected should the need arise. Provision is also made inside the panels for service cables such as telephones and electricity. Proprietary non-loadbearing partitions are also made of steel or aluminium sheet into which glass panels or doors can be fitted and interchanged in a standard frame. These are in popular demand for offices, industrial building, and stores.

12.3 Typical internal walls and openings

BRICK WALLS. The most solid and traditional of all internal walls is probably the 'one brick' wall often used as a party wall between buildings and for crosswall construction. Standard sizes of clay bricks are given in B.S. 3921. The designated clay brick format is 225 × 112.5 × 75 mm but to obtain the actual work size 10 mm must be deducted from each dimension (Fig. 12.1). The above dimensions do not apply to the metric modular brick also shown in Fig. 12.1.

The brick is usually solid, though perforated wirecuts are also used (Fig. 12.1). It is strong, fire-resisting, and still in wide use despite other systems of internal walling. It is 215 mm or one brick thick – hence its name. Apart from partitions it is also in demand for division or separating walls and as support for staircase steps and in buildings other than for domestic use. It is also used as a fire-break where fire regulations have to be met.

Half-brick walls are also a popular form of internal construction, particularly in buildings for industrial use, storage warehouses, or where strength and stability are needed (Figs. 12.1 and 12.2). They are capable of acting as loadbearing walls and can also support fairly heavy doors, sliding shutters, shelving, brackets, etc. They also act as cross-wall stiffeners in long buildings when built at reasonable intervals. For small or domestic work the compressive strength of the brick is usually ample as the lowest compressive stress of 5.2 N/mm^2, as laid down by B.S. 3921, is easily met. Brick walls, however, do require stable foundations and if started at an upper floor, need a beam or similar means of strengthening the supporting floor. In order to reduce the dead weight of the wall, perforated or deep-frogged bricks are commonly used (Fig. 12.1).

BLOCK WALLS. A *clay block* is defined as 'a walling unit exceeding in length, width, and height, the dimensions specified for bricks'. The designated clay

Internal walls: function and basic construction | 123

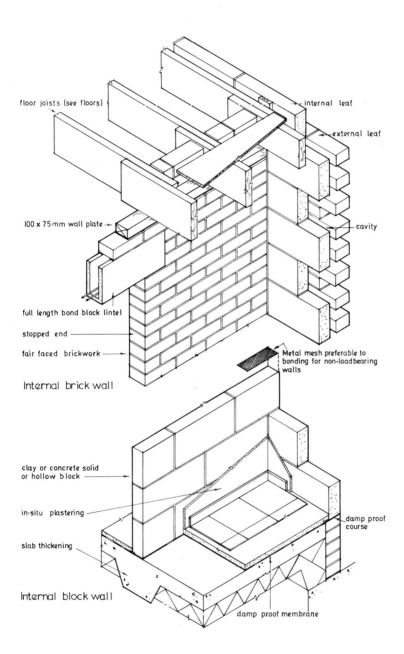

Fig. 12.2 . Internal walls

block comes in sizes of 300 mm long by 225 mm high and in widths of 62.5, 75, 100, and 150 mm, but for actual working sizes see Fig. 12.3. Burnt clay blocks are available in different styles and patterns and for internal work they can be supplied with a keyed face for plaster finish (Fig. 12.3).

A standard clay block 100 mm thick has a compressive strength of 4.5 N/mm^2 which more than covers the B.S. requirements of 2.8 N/mm^2. It is therefore loadbearing, whereas the clay partition block is not. Blocks should be checked against the imposed loads they will have to carry, but in most cases they are adequate.

Concrete blocks are used for both light and heavy internal work and are usually classified as *solid, hollow,* or *cellular,* i.e. with pockets in lieu of holes (Fig. 12.3). They are made in a range of sizes and are fairly easy to lay; also hollow, loadbearing blocks have the advantage of vertical spaces which can be filled with concrete, reinforced if necessary, to act as columns or internal supports. The cavities can also accommodate services in some instances. Ends of blocks can be supplied either square or tongued and grooved as shown for greater stability, or made with recesses to strengthen the mortar joints. Care must be taken, however, with hollow blocks which have to carry wall furniture or shelving, to ensure that they can take the load.

Non-loadbearing walls of lightweight concrete should always be made secure on all four boundaries to ensure rigidity. Where one or both intersecting walls are non-loadbearing, they should preferably be connected by metal mesh rather than by bonding to permit slight movement without cracking (Fig. 12.2). Tops of walls are secured either by battens (Fig. 13.4) or by slate in mortar packing at ceiling level to solid upper floors. In some industrial areas where coal is plentiful, breeze partition blocks are still in demand for partitions, though these are being superseded by other lightweight materials.

Concrete blocks, like clay, are usually specified by their co-ordinating or overall sizes as fixed, with the actual work size 10 mm less to allow for joints. The main sizes are shown in Fig. 12.3 but for the full range the reader is referred to B.S. 2028.

OPENINGS. Openings in internal masonry walls are not difficult to form. The width will depend on the purpose for which they are intended (see door/window openings, Chapter 8). The jambs may be square or rebated but are usually square for internal walls (Fig. 12.1). Masonry above the opening is usually supported by a lintel, which may be of timber, concrete, steel joists, angles, or reinforced masonry (Fig. 12.2). For masonry walls up to half brick thick, timber may be used for spans up to one metre. The exact depth may be calculated, but for a normal door-opening a timber lintel of about 100 mm wide and 75 mm deep is adequate. For larger spans a concrete lintel can be used, reinforced with m.s. bars placed near the bottom of the beam (Fig. 12.3). They are usually made of 1:2:4 concrete with one steel bar to each half brick for

Internal walls: function and basic construction | 125

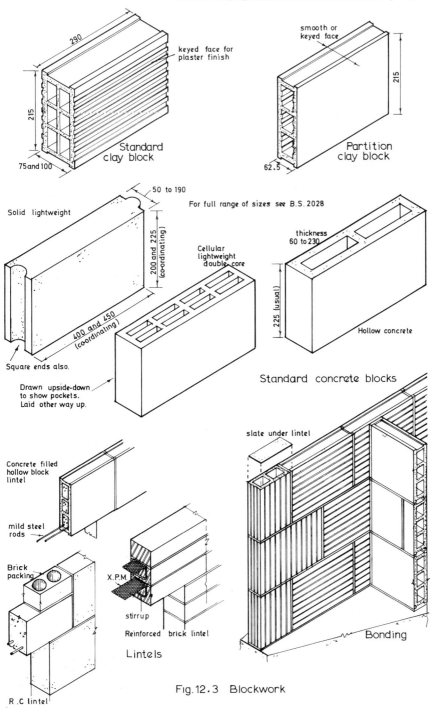

Fig. 12.3 Blockwork

most small openings, or two bars for wider spans as shown. Above this, the beam sizes and bars are usually calculated.

The lintel may be pre-cast or cast *in situ* whichever is convenient. It is advisable to make the depth of the lintel equal to a multiple of a brick course to avoid packing or cutting of bricks. The bearing on brickwork in small openings is usually about 108 mm to facilitate bonding, but for larger spans should be made equal to the depth of the lintel.

The lintel shown in Fig. 12.2 is a pre-cast U-shaped beam which is fairly light and can be lifted into position by hand. It can also be cast *in situ* by placing 400 or 450 mm blocks on bearers, inserting the reinforcement, and then filling the cavity with concrete.

Reinforced brickwork is also popular for lintels. One method of doing this is to carry the brickwork across the opening on a bearer and strengthen it by adding steel reinforcement of either m.s. bars or expanded metal mesh with stirrups as shown in Fig. 12.3.

In factory or storage building where strength is required, m.s. angles are sometimes used as lintel support for brick- or blockwork. A single angle is used for half brick walls or two angles, one each side, for a one brick wall. In situations where rough treatment can be expected, such angles can protect the head of the opening. For wide spans, steel joists of I section are used. Pressed-steel channels are also popular especially where concrete blocks can be laid between the flanges to give a neat finish.

12.4 Finishes to internal walls

The normal finish to internal brickwork is usually one of three kinds, *self-finish*, *wet-finish*, or *dry lining*.

Self-finish. In industrial and utility building, schools, and common parts of buildings including stairs, landings, stores, etc., it is usual to use fair-faced brickwork or blockwork with neat mortar-pointed joints. The wall may then be left in its natural state or coated with emulsion paint or similar finish. The practice of using fair-faced brickwork — or coloured concrete brick — internally has grown in recent years and is now a common finish in modern building both commercial or domestic. In stores, warehouses, and workshops it is customary to coat walls with hard gloss paint to provide a durable surface. Natural stone in random sizes is also used internally for features in modern building.

Wet-finish. This process is traditional and still in wide use, though not as much as formerly. It consists of smooth plaster finish often applied in two coats: the walls are first keyed to provide a grip and then rendered with an undercoat of plaster and finished with a smooth setting coat. Other plaster finishes are also available. Most plasters are of proprietary brands, delivered to site in 50 kg bags and are ready-mixed, only water needs to be added. Often the surface of the wall is grooved to provide a key (Figs. 12.1 and 12.3). Owing to the scarcity of plasterers and because buildings take some time to dry out, the wet process is now giving way to dry lining.

Dry lining. This consists of sheets of plasterboard or timber-based board secured to walls by sticking or nailing to grounds. The sheets are usually about 2400 × 1200 mm though other sizes are marketed. The joints can be covered in many ways; some form of cover strip or bevelled edge is usual for timber-based board, but plasterboard is normally dealt with by filling the joints with plaster to produce an invisible joint.

Walls may also be finished in tiles of clay, plastic, glass, etc. of all sizes and colours, or panelled in various ways, or treated with coloured plaster or other finish in a range of patterns. They can also be sprayed or painted. Wall finishes including partitions will be considered in more detail at Level 2.

13 | Floors: function and basic construction

13.1 Primary functions of ground floors

The fundamental purpose of a floor is to act as a flat, usually horizontal, supporting element of a structure. It is generally located by reference to its position or 'level' in the building. Since the earliest days of building, the structural design of the floor has undergone many changes and the variety of construction and materials used is now quite extensive. In this chapter, however, attention will be confined to ground floors of the solid and suspended timber types and upper floors of the suspended timber type.

Basically the purpose of the ground floor is to provide a firm, dry platform for people and things, i.e. furniture, stores, plant, and equipment, and its construction will vary according to the purpose for which it is to be used. As a structure itself is designed for a specific purpose — domestic, industrial, recreational, etc. — so the floor is also built to a particular need; any change of use later must usually be notified to the appropriate authority.

The main functional requirements of a floor are:

strength and stability,
damp exclusion,
thermal insulation,
fire-resistance,

though not necessarily in that order.

STRENGTH AND STABILITY. With solid ground slab floors the question of structural strength rarely arises, particularly if the floor is made of sound material such as concrete mixed in proper proportions and well laid on a firm base. In such cases and when subject only to pedestrian or light-wheeled traffic it will last indefinitely, though of course the durability of the surface will depend on what sort of floor-finish is used. Where there is any likelihood of movement in the subsoil due to shifting sand, expansion of clay, subsidence, etc., it is advisable to test the soil before building. This is now fairly common practice as explained in Chapter 5.

Suspended ground floors of timber are not as popular now as hitherto, and are normally used only where extensive fill would be necessary, or where the space below the floor is needed for a specific purpose. Such floors would be constructed over a basement area, on sharply sloping ground, or would be built on dwarf or sleeper walls (Fig. 13.1). The floors are constructed in a similar manner to suspended timber upper floors: suspended ground floors on joists or sleeper walls are usually much better, as the walls shorten the span thus giving greater rigidity. All suspended ground floors are now required to be insulated (Fig. 13.3).

DAMP EXCLUSION. Moisture penetration at floor level can be prevented provided necessary precautions are taken. Two methods are in common use for solid ground floors; the first is designed to prevent rising damp due to capillary action and is achieved by using standard damp-proof courses and membranes; the second, designed to keep the inside of the wall and the floor edge dry is done by using cavity walls. The combined use of the d.p.c., d.p.m., and cavity is reasonably effective in keeping the building free from damp.

Suspended ground floors are usually constructed on sleeper walls (Fig. 13.1). Here the floor joists and wall plate are protected from rising damp by a d.p.c. as before. Damp can bridge the cavity, however, if it should flood or contain mortar droppings; for this reason the cavity foundation fill must not reach to within 150 mm of the inner leaf d.p.c. The ends of the floor joists, too, should avoid contact with the inner wall, a precaution which also permits some expansion of the floor if necessary.

THERMAL INSULATION. Thermal insulation now plays a much more important part in building construction than hitherto and thermal properties of materials have received close attention at the Building Research Establishment. The use of foamed and expanded plastics for thermal insulation to floors — as well as walls and roofs — has increased greatly in recent years. It is for this reason that suspended ground floors are being superseded by the solid ground floor. Owing to underfloor ventilation, the heat loss of the former can be considerable and some form of insulation beneath the boarding is usually insisted upon (Fig. 13.3). But even with the solid floor the heat loss can be quite high as a result of the cold bridge between the external wall and the ground floor slab, so precautions are taken in some types of industrial systems to curb heat loss (Fig. 13.2).

FIRE-RESISTANCE. The degree of fire-resistance provided by a solid ground floor is normally adequate. With suspended timber ground floors, including those covering basements, the degree of fire-resistance will depend largely on user requirements. These and other factors will be considered at a higher level.

13.2 Solid ground floors and external walls

A typical sketch of a junction between a solid ground floor and the inner leaf of an external cavity wall is given in Fig. 12.2. The ground floor in such cases is normally about 150 mm above the outside ground level (Fig. 13.1C), but not necessarily so (Fig. 13.1B). This external height of d.p.c. above the ground

130 | *Construction Technology*

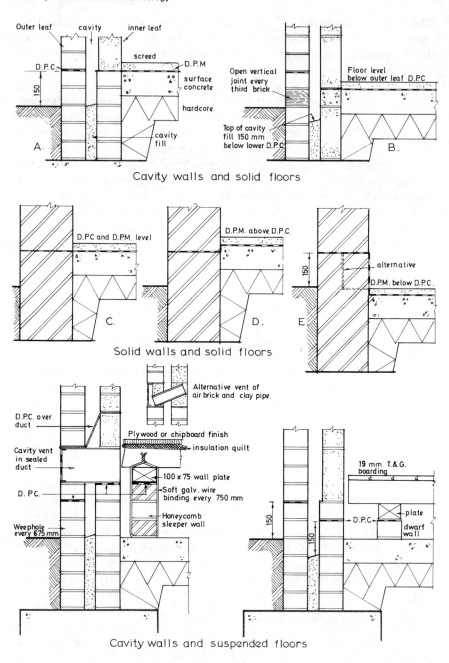

Fig. 13.1 Junction between ground floors and walls

level is needed to ensure that surface water does not flood or splash up above the d.p.c. or that any small change in the external ground level due to the garden soil, paths, rubbish, etc., can be accommodated. Usually the base of the cavity is drained at intervals of three or four bricks by leaving weepholes in the vertical brick joints (Fig. 13.1).

In order to keep the wall and ground floor dry, it is not sufficient merely to insert a horizontal d.p.c. in the wall at 150 mm above the ground; solid ground floors cannot always be constructed to achieve this, particularly where cut and fill of soil is necessary, Chapter 5. The position of the ground floor relative to the ground is therefore important with solid walls and vertical d.p.c.s may be necessary (Fig. 13.1E.). Where cavity walls are used in such cases, the vertical d.p.c. may be dispensed with (Fig. 13.1B). Here the inner horizontal d.p.c. is lowered to suit the floor and the top of the cavity fill adjusted accordingly.

When considering foundations in Chapter 6 it was shown that for smaller structures a floor slab thickened at the edges was used, and for certain types of construction and in lightweight industrial systems this is fairly common practice (Fig. 6.12). Another system using edge-beams is given in Fig. 6.5 and the relationship of the foundation to the solid floor is also shown in Fig. 13.2. Edge insulation, usually of expanded polystyrene between 19 and 25 mm thick is almost always used where the floor is artificially heated. Even when this is not so it is now customary to insert vertical insulation as shown.

It should be noted that a damp-proof membrane is not always necessary in solid ground floors and is often omitted. This is usual where site conditions are fairly dry, where no ground floor heating is installed, or where impervious floor finishes are used.

13.3 Hardcore

The foundation surface concrete or the solid ground floor can be affected by the kind of soil on the site and the load it will safely carry. The first site operation is usually the removal of vegetable soil to a depth of about 150 to 250 mm. Unless the ground is consistently firm and likely to remain so throughout the year, it is normal to lay surface slabs on a bed of *hardcore.* This consists of broken brick, stone, or concrete rubble, which should be clean, hard, and free from soft mortar, dirt, or rubbish, and is usually laid about 100 to 150 mm thick. The object of this is to form a firm dry base for the concrete slab. The thickness of the hardcore is often increased to suit the finished floor level or surface concrete, in which case it is laid in layers of 100 to 150 mm thick, each layer well compacted.

Materials for hardcore should be inert, i.e. contain no substance which could attack concrete such as colliery waste, gypsum, or any material containing sulphate. It must not expand or be otherwise affected by moisture. After laying

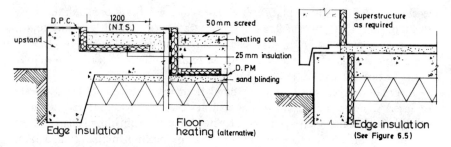

Fig. 13.2 Concrete foundations and solid floors

it is usually rolled or tamped to a minimum thickness of 100 mm and then *blinded* with a thin layer of sand or inert ashes. Should the material used as hardcore be suspect, then a layer of polythene or other waterproof sheeting would be placed over it before the concrete is laid.

Hardcore is very useful on muddy sites for keeping the surface concrete clean and also for reducing the capillary action which could cause rising damp. Small-size hardcore can also be laid in thin layers as a 'blinding' to keep the site concrete clean. Sometimes it is mixed with cement to form a weak concrete carpet 1:12, and laid 50 to 75 mm thick. This carpet is frequently specified as compulsory in the event of over-excavation when soil is not permitted to be returned as fill to make up levels. An example is shown in Fig. 6.9.

13.4 Suspended timber ground floors and external walls

Typical sections of junctions between cavity walls and suspended timber ground floors are given in Fig. 13.1. As with solid ground floors, the external d.p.c. is placed at a minimum distance of 150 mm above the ground. It will be noted that the suspended ground floor is supported by sleeper walls which can be built to any required height, thus eliminating the need for extensive ground fill. The ground floor is also out of contact with the ground. The sleeper walls are normally built directly on the surface concrete, which in turn is laid on hardcore as just described. The hardcore is normally at 'site strip' level so that the top of the surface concrete is above ground level. It is not then liable to act as a sump for surface water.

The sleeper or dwarf wall is normally half brick thick and honeycombed to allow free air circulation beneath the floor. The floor joists are normally supported by timber wall plates to which they are nailed. The plate rests on a d.p.c., usually of bitumen felt supplied in rolls of 100 mm thick, which prevents rising moisture from the wall reaching the timber.

In order to obviate the necessity for building the ends of joists into external walls, it is usual to erect sleeper walls next to the main walls and about 50 mm from them. Where this is impracticable, the ends of the timber should always

Floors: function and basic construction | 133

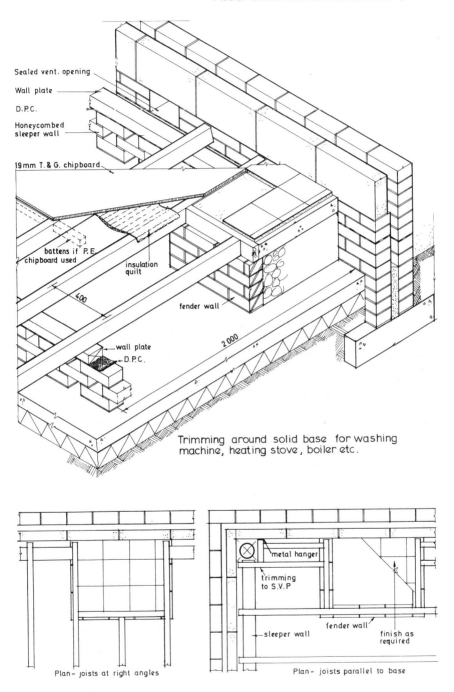

Fig.13.3 Joist trimming — suspended ground floors

be treated with preservative to prevent rot. Where suspended ground floors are used, an insulation quilt is usually required beneath the boarding to prevent heat loss (Fig. 13.3).

The space beneath the floor must be adequately ventilated in order to prevent fungal growth of which dry rot is the most common variety. As this thrives in unventilated, humid conditions, it is usual to insert air bricks or ventilating pipes around the external wall (Fig. 13.1). These are sealed off from the main cavity and are spaced in the ratio of 968 mm^2 of open ventilation space to every 305 mm of wall. The ventilating grilles are normally made to suit brick sizes, either one or two courses high.

13.5 Comparisons between solid and suspended ground floors

The suspended timber ground floor has the advantage of light weight and is of dry construction. It is fairly simple to construct and economical where loads are not heavy. Sleeper walls are easily built up to the required height without the need for extensive filling or making up of levels. Joists may be spaced to suit the type of timber-based board flooring where this is used. Openings are readily formed to allow for solid bases for fireplaces, ducts, or other projections, and pipes, cables, and services are easily secured to joists or laid in the underfloor space. The suspended floor is much in favour with the small builder who, unlike the builders involved in large projects like housing estates, may lack equipment to move, level, and ram large amounts of soil or fill.

This sort of floor has several disadvantages, however. It can be relatively expensive in timber costs, it is of low fire resistance, its thermal insulation value is not high, and it needs to be adequately ventilated beneath the floorboarding to reduce the incidence of dry rot. It does not possess the rigidity or strength of a solid floor and should it shrink or settle around the walls, the possibility of draught is greatly increased. The outside vents are also liable to blockage from rubbish or garden soil, which either arrives there accidently or is purposely inserted to cut down draughts. Also, as surface concrete is essential beneath the floor, this tends to reduce any economic advantage it might otherwise have.

The concrete solid floor has the advantage of strength and fire resistance. It can be laid conveniently when the foundations are at d.p.c. height, thus providing a good working platform for building the superstructure. The floor can be protected against damp by insertion of a d.p.m. and a much wider range of finishes is possible than with a timber-joisted floor. A hard-wearing surface can be added to suit most purposes; it is more durable and better suited to resist vibration than a suspended floor. Strip foundations for internal load-bearing walls can easily be provided by thickening the floor slab in most cases.

One disadvantage of the solid floor is the difficulty of dealing with services. Floor wells and recesses must be accurately located or pipes and cable laid within the screed thickness and secured while the floor concrete is green. Omissions are difficult to remedy, accessibility is not always easy, and additional services sometimes expensive to install. Impact noise and vibration from traffic can also be greater than with suspended floors.

In situ floors are usually of 'wet' construction which may entail a drying-out period. Provision of fill to reach required floor levels can sometimes be costly and inconvenient; cleaning and hacking the floor for screeding later can be time-consuming unless a proprietary adhesive is used or the screed laid when the concrete is green. An untreated surface usually results in the screed lifting. However, because of the increasing cost of timber and the widespread use of concrete-mixing and earth-moving equipment, solid floors are now in wider use than suspended ones, especially where estates are being built.

13.6 Suspended timber upper floors

The standard arrangement with timber upper floors is for common or bridging joists to span between walls or other supports with the ends of the joists held by metal hangers or bearers built into the wall. The size of the floor joists will depend on the spacing, bridging span, and the load they will have to carry. Tables of acceptable sizes to suit different spans and loading may be found in current Building Regulations, Part D: Schedule 6, which also gives rules for determining dimensions. Rule-of-thumb methods for arriving at depths of joists should not be used, particularly as the width of the joist can vary. Provided the reader has an adequate knowledge of structural mechanics he can calculate his own dimensions.

Joists are normally spaced at 400 mm centres in order to suit both the width of the chipboard or plywood flooring sheets but also to provide bearing for the ceiling which usually consists of plasterboard (see floor finishes). Ends of joists can either be built into the wall on felt or slate packing (Fig. 13.5) or supported by metal brackets or joist hangers (Figs. 13.4 and 13.5). Hangers are to be preferred as they are speedy, save the necessity for pockets to take the joists, and are particularly useful with industrial concrete building panels of storey height.

Where openings have to be formed in the upper floors to allow for fireplaces, staircases, service ducts, etc., the normal practice is to *trim* the opening (Figs. 13.4 and 13.5). In order to construct the opening, *trimming, trimmer,* or *trimmed* joists are used. The trimming joists bridge the whole span from wall to wall; the trimmers are supported at one or both ends by the trimming joists; the trimmed joist is supported at one or both ends by trimmers (Fig. 13.5). The trimming joists and trimmers are made thicker than the normal floor joists; sizes can be obtained from the building regulations quoted above, and can be up to about twice the normal thickness depending on the load (Fig. 13.4).

136 | *Construction Technology*

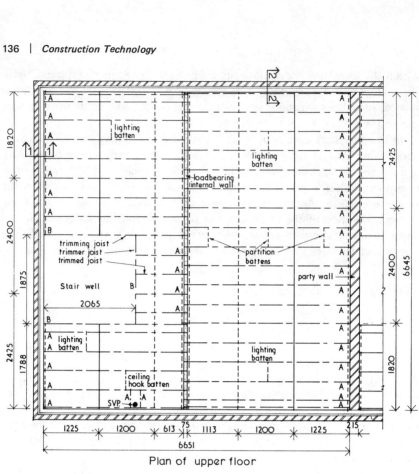

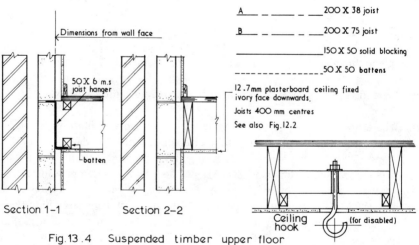

Fig. 13.4 Suspended timber upper floor

JOINTS. There are various methods of securing one joist to another and the traditional mortice and tenon joint in carpentry has been in use for centuries. The best known of these is the *tusk tenon* (Fig. 13.6) used to secure the trimmer to the trimming joist. Though this is still used to some extent, it has now been largely superseded by metal hangers or patent pressed-steel connectors, especially for dwellings or where loads are not excessive (Fig. 13.5). Pressed-steel connectors or framing anchors are preferred to the square housing joint (Fig. 13.6) and are also used in roofs, partitions, and timber-framed constructions.

Building regulations specifically state that ends of joists must not be built into any part of a fireplace, hearth, or flue, nor shall timber fixing plugs be used: in such cases it is usual to use metal brackets as shown in Fig. 13.6. Upper floors are normally required to cover the whole span in one length without splicing. Overlapping, however, is commonly used on loadbearing walls or supports (Figs. 12.2 and 13.4).

13.7 Strutting suspended timber upper floors

Where joists have to be deep in relation to their width to be strong enough to cover the required span, it is necessary to provide stiffeners at intervals to prevent winding or buckling. This is normally done by strutting, and the two usual methods are *herring-bone* and *solid*, the latter being either *staggered* or *in-line* (Fig. 13.5). There are certain official recommendations made in respect of size and spacing which will be considered at another level; the principle, however, can be seen in Fig. 13.5. Solid strutting is simple and quick but has some disadvantages. Where the joist is slightly warped, for instance, it sometimes fails to make good contact throughout its full depth. To overcome this, a metal rod with threaded ends is often passed through the joist near the strutting and tightened up as shown.

A better method is to use herring-bone strutting, where each strut acts as a brace giving positive support to the top and bottom of each joist. By inserting folding wedges between the end joist and the wall, the joist can be stiffened up quite firmly. Another advantage of herring-bone strutting is that the space between the joists is left partly clear enabling service-pipes and cables to be installed without difficulty. But the advantage of *in-line* solid blocking or strutting is that in domestic upper floors it is usually possible to position the struts to act as battens to which the edge of the floor and ceiling sheets can be secured (Fig. 13.4).

13.8 Floor finishes in domestic construction

Available floor finishes for domestic use are very numerous. Some are suitable only for solid floors, others for suspended timber, but most can be used for both. For suspended ground floors, tongued and grooved or plain-edged board-

138 | *Construction Technology*

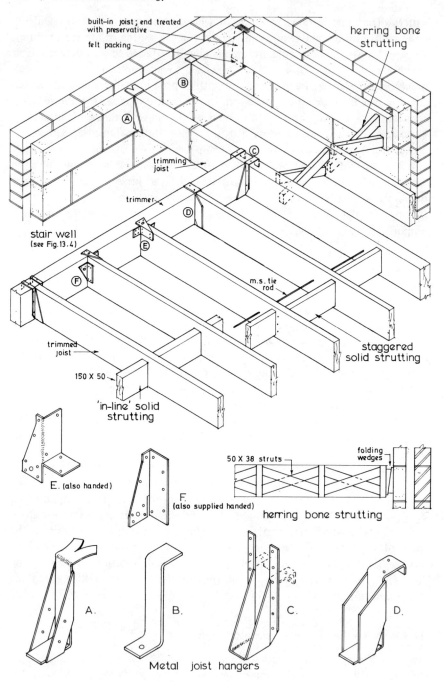

Fig. 13.5 Trimming and strutting – suspended upper floor

Floors: function and basic construction | 139

ing, or timber-based sheets are usual. Smaller builders still use t. and g. but contractors engaged on multiple or repetitive units prefer chipboard or plywood sheets for speed and economy, and the use of traditional flooring is now in decline.

Finishes to solid ground floors are normally applied to a levelling screed of 38 mm cement and sand, either floated or trowelled smooth to receive the finish. The choice will depend on a number of factors such as cost, durability, colour, hardness, noise absorption, ease of maintenance, or non-slip properties. Domestic finishes can be divided into four broad categories:

        jointless   slab
        sheet   wood

JOINTLESS. The most common of these is the cement and sand screed consisting of one part cement to three parts sand and laid about 38 mm thick on a clean and roughened surface. The finish is usually trowelled smooth to receive carpet or sheet flooring.

Granolithic concrete finish is more durable. It consists of one part cement to three parts granite chippings laid in the same way as cement and sand. It is used for paved areas, yards, stores, and garages, etc., which are subject to

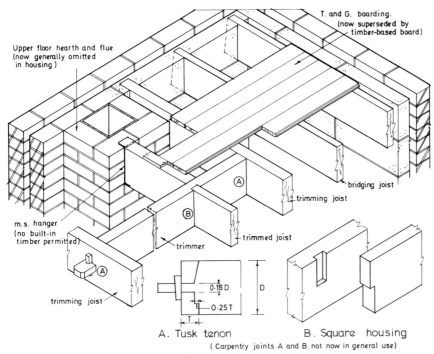

Fig.13.6 Trimming to upper floor joists

fairly heavy traffic, but it is not usually used for domestic work.

Terrazzo is popular for entrance porches, kitchens, and small external areas, also in communal areas and stairs to blocks of flats. It consists of two parts marble chippings of various colours to one part Portland cement laid *in situ* about 13 mm thick on a cement and sand screed. After the surface has set hard, it is ground smooth by machine. Synthetic resins or plastics are also available in paste form for applying to the screeded floor. Terrazzo is now mostly supplied in tile form for domestic work.

SHEET FLOORS. Linoleum, cork, and plastic are commonly used as coverings either on suspended or solid floors. PVC (polyvinyl chloride) flooring takes several forms; it can be supplied in rolls about 25 metres long and one or two metres wide, stuck down with strong adhesive if necessary. Cork flooring to bathrooms and kitchens is normally stuck down, but other forms of sheeting are often laid loose.

PVC and vinyl tiles are light, cheap, and produced in a range of colours. The screed is usually coated with primer and tiles stuck down with an adhesive. As with all forms of thin sheet flooring it is essential that the screed is trowelled smooth to prevent irregularities from showing through.

SLAB FLOORS. Quarry and vitreous tiles are popular in kitchens, porches, and external areas. The vitreous types are usually of better quality and can be produced in a variety of bright colours and patterns; the surface can also be rippled to produce a non-slip finish. Concrete and terrazzo tiles also make a durable finish and are available in several shapes and colours. The practice of using curved-edged, vitreous interlocking tiles in continental jigsaw pattern is also growing, and many attractive styles and designs are now on sale.

WOOD. Good quality hardwood of highly polished strip is sometimes laid in living rooms in lieu of sheet-flooring or carpet. This can be secret-nailed to joists or to bearers secured to solid floors with floor clips. This type of finish, however, is mainly confined to small community buildings which may be used occasionally for dancing.

Wood mosaic panels are cheaper than hardwood strip and are now popular. They are easy to lay on either screeds or boards and are supplied as large-size wood tiles about 450 mm square, usually in herring-bone or basket-weave pattern and stuck down with adhesive. Other wood-floor finishes are also in use and will be considered at another level.

Timber-based sheets, usually of chipboard or plywood, are now the most popular type of finish for domestic and similar small building. They consist mostly of 19 mm tongued and grooved or 22 mm plain-edged boards (Fig. 13.3). Where p.e. boards are used it is necessary to support the ends on battens placed across the joist (Fig. 13.3). Sheet sizes vary but the usual size is about 2400 ×

1200 mm. It is important that only one size of sheet is used; that plasterboard ceiling and the floor sheets are of the same size; that this is decided before the joists are fixed; and that accurate spacing is maintained (Fig. 13.4).

It will be seen that both floor and plasterboard sheets are laid lengthwise across the joists. As internal dimensions are normally taken from the wall face, Section 1 – 1, allowance has to be made for this when spacing the joists – 25 mm in this case. Solid strutting or blocking is used and fixed 'in-line' to provide edge support for the plasterboard and also for p.e. floor-sheets where these are required. Both plasterboard and floor-sheets should be trimmed to size so that the cut edge butts against the wall face and not over the joists or noggins, otherwise plasterboard particularly cannot be properly jointed. Figure 13.4 allows for this. All noggins for bearers, ceiling lights, and special fixings must be inserted before the floor and ceiling sheets are fixed; also non-loadbearing partitions need to be supported at the top edge and battens are inserted for this purpose.

14 | Stairs: function and basic construction

14.1 Function of a stair

The stair, when considered as part of the structure, is as much a functional element as the wall or frame. The term 'staircase' originally referred to the space or case in which the stairs were built. The terms 'stairs' and 'staircase' are now used indiscriminately, although the latter is still used to describe the enclosure, especially in modern buildings, which sometimes use glass walls or open screens to house the stairs. Generally speaking, the *stair* is one of a series of steps connecting two floors while a flight is a series of steps between landings or platforms.

When designing stairs for a specific function it is essential to consider the basic purpose, i.e. to give access from floor to floor and to provide escape in the event of fire, both of which have to be examined within the framework of certain criteria. These basic requirements for stairs are:

strength and stability,
sound-insulation,
fire-resistance,
appearance,
feasibility.

STRENGTH AND STABILITY. Clearly the strength of a stair will depend on the purpose for which it is intended. For buildings other than domestic housing of one or two storeys, the imposed load, i.e. the people and goods it will have to carry, is normally laid down or recommended in official Codes of Practice. Sometimes structural calculations are made to ensure that the architectural design conforms to specific requirements. These are not usually demanded at domestic level but even the most straightforward stair has to meet certain standards.

Stairs can be built of reinforced concrete, steel, or timber or any combination of these. In addition to the stair itself the question of strength and stability of supporting walls and floors to fire-exits is also important. Generally, the narrower the width of the stair the greater the strength.

SOUND-INSULATION. In some buildings the problem of airborne and impact noise due to stair construction can be troublesome and preventing the passage of sound from one floor to another is often of primary importance. Hard-wearing surfaces on stairs and landings can give out impact noise when walked upon. Where staircases are isolated from the main structure by closed doors on landings, as is usual in public buildings, noise transmission is considerably reduced. But internal stairs, whether for industrial, commercial, or domestic use do allow airborne sound to travel freely, even when the surface is of sound-absorbent material. The degree of sound-insulation required will depend on the purpose of the structure.

FIRE-RESISTANCE. The importance of the staircase as a fire escape route has already been emphasized. The construction of the steps themselves is as important as the supports, and the materials must also have a measured degree of fire-resistance, which will be discussed at another level. In the small-scale and domestic buildings with which this volume is concerned the question is not quite so serious.

APPEARANCE. As an architectural feature in a public building the staircase was once ranked as of first importance and designed usually as a central feature. With the advent of the lift, however, and the need for space-saving and economy, the staircase is now usually only functional, and appearance tends to be of secondary importance, though there are exceptions. In domestic work the tendency is to make the staircase as compact and economical as buildings regulations will allow.

FEASIBILITY. A staircase must be constructed in such a way that it is safe and comfortable to use. The width, pitch, and height of the stairs must be confined to practical limits, as must the number of steps between resting-places and landings. Handrails, balustrades, and steps must be safe and of reasonable height, and the needs of children must be catered for. Such requirements vary according to use and location, and the need for economy in terms of space as well as cost must be considered. The minimum requirements for stair design in dwellings are laid down in building regulations as given in Fig. 14.2.

14.2 Constituent parts of a straight-flight stair

The elemental components of a straight-flight stair may be identified as follows:

Staircase. This has been described in the introduction. In dwellings a *common stairway* is one used by more than one person or family whereas a *private stairway* is used only by a single family or unit.

Step. Actually this is the horizontal surface of a stair. In practice it is often referred to as a combination of:

tread, i.e. the horizontal surface of a step, and

riser, i.e. the upright (vertical or near vertical) surface of the step. The term should not be confused with *rise* which is the actual height between treads (Fig. 14.2). In some staircase construction the riser is omitted to give an

open-tread stair (Fig. 14.1).

144 | Construction Technology

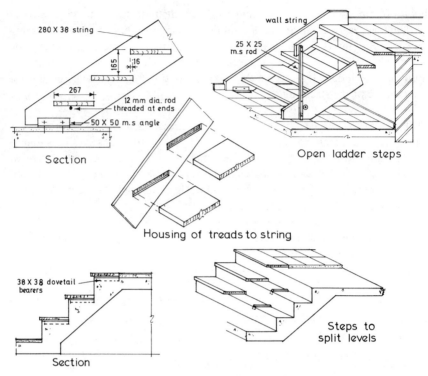

Fig.14.1   Short flight steps

Newel. This is the vertical post placed at the top and bottom of a stair flight to support strings, handrail, balusters, and bearers as necessary (Figs. 11.2 and 14.5). It is also the central pillar of a spiral stair (Fig. 11.1).

Nosing. This is the projecting edge of the tread, often rounded, formed at the external junction of the tread and riser (Fig. 14.3). The term is also used for the projecting, round-edged board to the edge of landings, etc. (Fig. 14.4B). A 19 mm projection is usual, as anything greater could cause tripping.

Flier. This is the term used in straight-flight stairs where the steps are rectangular on plan and of equal width throughout.

Flight. This is a continuous series of steps from one floor or landing to the next. A straight flight is one which has no landings.

Landing. This is the horizontal platform at the termination of one or more flights of stairs (Fig. 11.2).

Balustrade. This is a row of *balusters* or small vertical pillars supporting a stair- or handrail, though the balustrade is now often made as a solid vertical infilling between the string and the handrail (Fig. 14.3). Parallel rails as shown in Fig. 14.5 are also in wide use. Some authorities insist that, in domestic work, the space between the intermediate rails is not more than 100 mm, particularly in housing.

Stairs: function and basic construction | 145

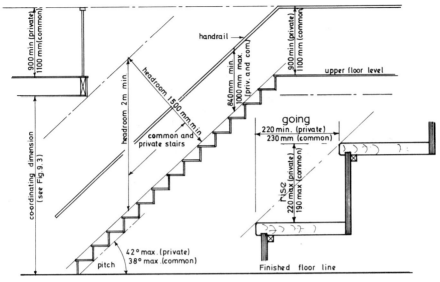

Fig.14.2 Private and common stairways (dwellings)

String. This is the inclined member or beam into which the tread and riser (if any) are housed. The inner string is usually, though not always, secured to the wall and the outer string is the independent member supporting the outer edge of the steps.

14.3 Critical dimensions in stair construction

Building regulations covering stair design or construction lay down critical dimensional requirements which must be adhered to (Fig. 14.2). These vary according to stair size and location. Dimensions for private or domestic stairs are not so stringent as those for common stairways. In straight flights critical dimensions apply to:

Rise. This is the vertical distance from the face of the tread to the face of the tread. Common stairways as in housing blocks must have a rise of not more than 190 mm, but stairs for domestic and private use can be up to 220 mm high (Fig. 14.2). The rise of a flight is the vertical distance from one floor landing to the next.

Going. This is the horizontal distance from the face of the riser to the face of the riser. Common stairways must have a going of not less than 230 mm but for domestic stairs this may be reduced to 220 mm.

Handrail height. The height of a stair balustrade, i.e. the vertical distance of the handrail measured from the line of the nosings must not be less than 840 mm and not more than one metre, for both private and common stairs. Private stair handrails to landings are not to be less than 900 mm, but for common stairs this is increased to 1.1 m (Fig. 14.2).

Headroom. The headroom of straight-flight stairs must be not less than 2 m measured vertically from the line of the nosings to the edge of the landing (Fig. 14.2).

Pitch. This is the angle of the slope of the stairs to the horizontal. For private stairs the pitch must not be greater than 42°, or 38° for a common stairs. Stair pitch is important in setting out as it can affect both the going and the headroom.

All the above dimensions must be regarded as critical and not necessarily those used in all stair construction. The maximum permitted steepness (pitch) for example, may not be suitable for the elderly or disabled. Setting out will be considered further in Level 2.

14.4 Construction of a straight-flight stair

This is the simplest form of stair and one which has no landing. (Upper floors, however, if of small area at the top of the stair, are often referred to as landings.) Building regulations permit a maximum of sixteen steps in a flight, but such a stair could be tiring to climb and twelve or thirteen steps is usually considered as a maximum (Fig. 14.4).

Modern staircases can be constructed both with and without risers. The former is known as the *closed riser,* which has both riser and tread and the latter as the *open riser* which has treads only. The open riser, formerly used only for step ladders or in stores, warehouses, and lofts is now in general use for all types of stairs (Fig. 14.1).

In internal construction it is also usual to see short flights of not less than three steps (Fig. 14.1). These are in frequent use in common entrance spaces or at split levels to ground floors. They can be either open or closed risers but are often of concrete with hardwood treads secured to fixing blocks in concrete (Fig. 14.1). Such short flights sometimes terminate in landings from which straight flights can be built.

CLOSED RISER STAIRS. Figure 14.4 shows a typical straight-flight stair of the type popular in domestic housing or small offices. It is usually of empirical design, its members not having been calculated for size but made to certain minimum standards.

The total height from floor to floor level must be known before the stairs can be set out. First, the co-ordinating dimension must be known if it applies (Fig. 9.3). To this must be added the depth of the joists as required by building regulations depending on the floor span. To this is added the depth of the floor and ceiling finishes. It is usual to consider the co-ordinating dimensions first as these could affect the door-frame heights (Fig. 14.4).

The floor plan must be studied on both ground and upper floor and position of doors and windows noted, together with partitions, openings, finishes, etc.

From this information the stair space can be decided and pitch and dimensions worked out. The construction details of the stairs can then be prepared, examples of which are given in Figs. 14.3 and 14.5. Other types of stair will be considered at Level 2.

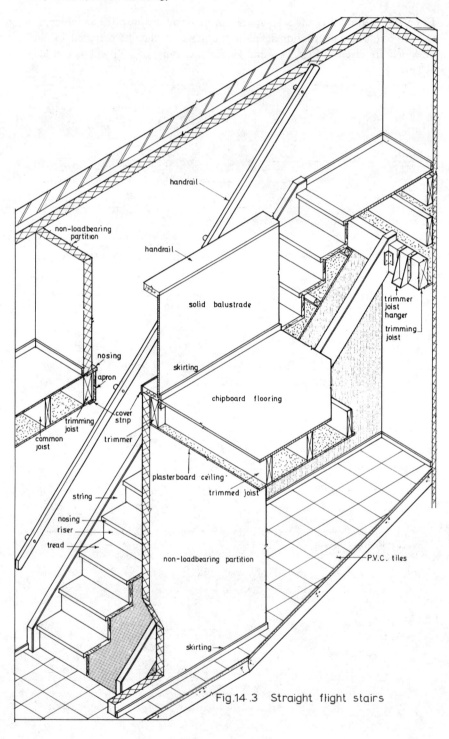

Fig.14.3 Straight flight stairs

Stairs: function and basic construction | 149

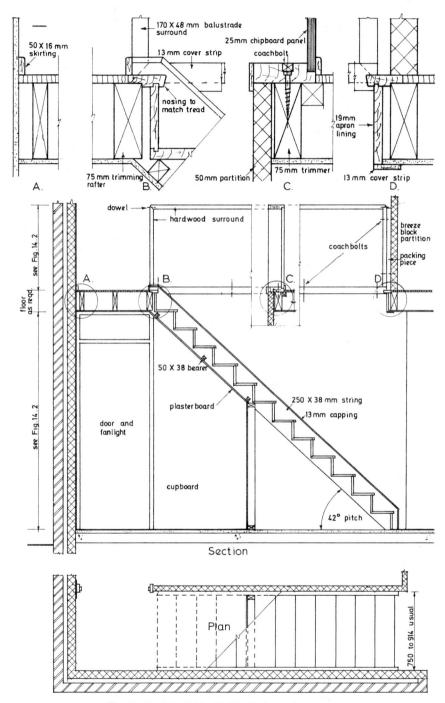

Fig. 14.4 Straight flight timber stairs

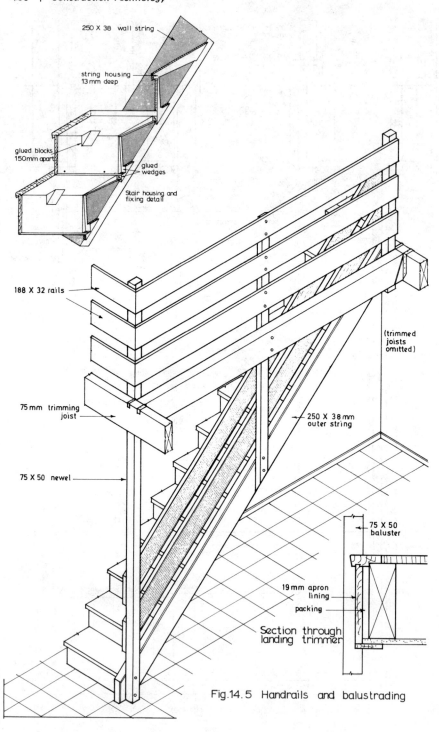

Fig.14.5 Handrails and balustrading

E. Services and external works

15 | Principal service installations

15.1 Basic requirements for water, gas, electricity, telephone, and drainage

Owing to technical research and development, improved methodology, competition among manufacturers, and the demand for a better standard of living, the provision of equipment and services has now reached a very high level compared with a few decades ago. The range now provided offers not only increased convenience for the consumer but also simpler installation and a higher standard of safety. In addition to the main services covered in this chapter, there are others available which can be installed even in the most modest of buildings. These include alarm bells, electric clocks, radio and television relay systems, door-bells, and phone entry-systems; but only water, gas, electricity, telephone, and drainage will be considered here. The primary requirements for these services can be listed broadly as being:

suitable for their purpose;
easily and neatly installed; and
reliable, safe, and economical in use.

WATER. The basic essential for water is that it be pure and fit for drinking. This is a normal stipulation whether the water is to be used for consumption or for industrial purposes. By the time it has reached the consumer it has usually been treated either by filtration, i.e. by passing it through sand beds or by adding chemicals to remove suspended matter. It will also have to be *sterilized*, usually done by adding chlorine in very small quantities. A third treatment, not always carried out, is that of *softening*. Hard water, though not injurious to health, has several disadvantages; it causes hot pipes and kettles to 'fur' by depositing lime-scale and also prevents soap from lathering. Both water softeners and filters may be installed by consumers if required, particularly if private supplies are in use.

Most buildings are supplied with water from public mains, and to maintain a uniform pressure an adequate 'head' of water is needed. This is normally done by constructing service reservoirs at convenient high points in the district and distributing supplies through trunk and street mains (Fig. 15.1). The consumer is then able to adjust the pressure at the main by using a stopcock or valve placed at the entry to the building, beneath the kitchen sink, or in both places. In most towns public authorities insist on the installation of a storage tank at the time of building, which is usually placed either in the roof space or in the airing cupboard. This provides a reserve supply against failure of the main supply and also reduces water pressure to hot water cisterns and bathroom appliances, w.c.s etc. Stored water should not be used for drinking.

GAS. Gas supplied to domestic consumers for heating, cooking, and hot-water supply is now usually natural gas. The main source of supply in Britain is from beneath the North Sea from where it is distributed by grid all over the country. It is about 90 per cent methane and has a high calorific value. Once the supply well has been sunk it is cheap to produce, the main costs being for purification and distribution. Natural gas is also supplied extensively to industry.

The main requirements of gas are that it be competitive with other fuels, safe in use, supplied at constant pressure and quality, and clean and convenient in operation. Where natural gas is not available, bottled *butane* or *propane* is widely used, supplied as a liquefied gas under pressure. Propane has a higher rate of evaporation from its liquid form making it suitable for storage in large outdoor tanks. Butane is normally supplied in cylinders. The gases are produced initially at refineries where crude oil is refined for petrol: they are odourless but a stenching agent is added for safety reasons. Their properties are similar to those of natural gas.

ELECTRICITY. The basic requirements are that it should be safe in use and supplied at constant voltage and adequate current to the appliances it serves. Though not always as economical as other fuels, it is usually highly efficient and adaptable, and has a wider range of uses than any other source of energy. As the risk of electrocution and fire must be reduced to a minimum, certain precautions must be taken which are compulsory and which are now standard practice. These cover insulation, voltage, protection against overloading of circuits, and earthing of appliances. Precautions are also necessary where damp conditions are likely to be encountered.

TELEPHONE. Telephone and telecommunication services, when installed in commercial and industrial buildings, must satisfy the electricity supply authority before connection can be made. The Institution of Electrical Engineers has issued a publication entitled *Wiring regulations for building* which covers materials, design, and workmanship. B.S. 1710 also deals with telephone cables and ducts to meet minimum requirements.

DRAINAGE. A basic essential for pipe systems of drainage is that they accept and remove waste and soil flow from buildings completely and discharge them into a sewer or treatment plant without delay. The system must be watertight, self-cleansing, and have entry-points closed by water seals which must not be broken by fluctuating air pressure. The system must be ventilated in such a way as to prevent nuisance or danger to health. The primary principles to be

Principal service installations | 155

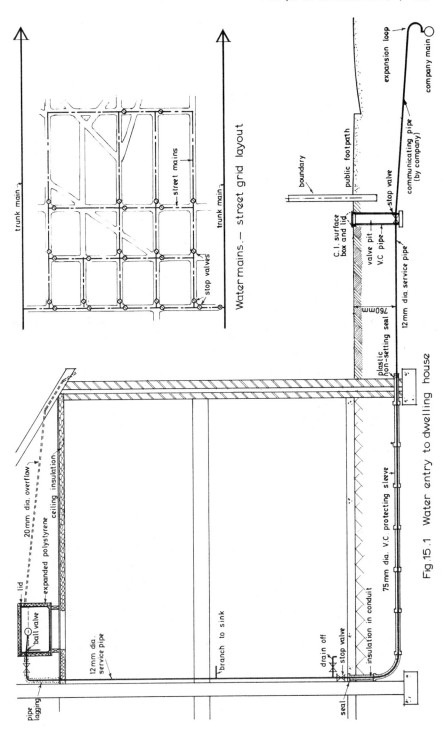

Fig. 15.1 Water entry to dwelling house

followed in designing and executing drainage layouts remain constant, and standard pipe systems are now in use everywhere.

As the local authority will normally be responsible for receiving foul water discharge, it is essential that its general requirements be met. Most authorities insist that separate systems of drainage be provided for foul water and rainwater, though some do permit *combined* systems whereby both may be collected in the same drain.

15.2 Provision in substructure for entry and outlet

Before deciding on the provision of services to individual buildings it is usual to contact local authorities and other appropriate bodies for details of existing services. A built-up area such as a town, industrial complex, or housing estate may already have a network or grid layout of basic services. These would include water, gas, electricity, street lighting, sewerage, and telephones, and an application to connect will have to be made. New developments would have service grids and networks decided at the planning stage and would be worked out mutually by the architect and the local authority. Before work commences, the site agent must be furnished with a ground-floor plan of the structure giving position of all services to be installed. Figure 3.14 gives part of a house plan which shows a typical arrangement.

WATER. Trunk water-mains in suburban areas are usually placed under the service road or footpath. The arrangement for town water supply is usually on the grid principle whereby two trunk mains are used which permit single street mains to be taken out of service if necessary (Fig. 15.1). The water company then provides a stop valve near the boundary of the new premises which is usually placed in a pit, often a vertical length of VC pipe with a cover at ground level. The stop valve is connected to the street main by a *communication* pipe and this is usually done by the water company or authority. A service pipe is then taken from the stop valve by the builder after the building substructure and carcass work has started. It is put in as soon as possible in order to provide water for the construction.

For domestic work the communication and service pipes are 12 mm diameter. A stop valve and drain cock must be inserted in the service pipe after it enters the building as shown. Service pipes must not be built into external walls: they should enter 0.76 m below ground and usually rise away from the external wall in built-in drain-pipes installed at the time of laying the foundation and ground floor slab. Figure 15.1 shows the general arrangement of connecting to the water main, and the position of entry can be shown on the setting-out plan (see Fig. 3.14).

GAS. Gas is usually piped directly from the grid and taken by service pipe to

Principal service installations | 157

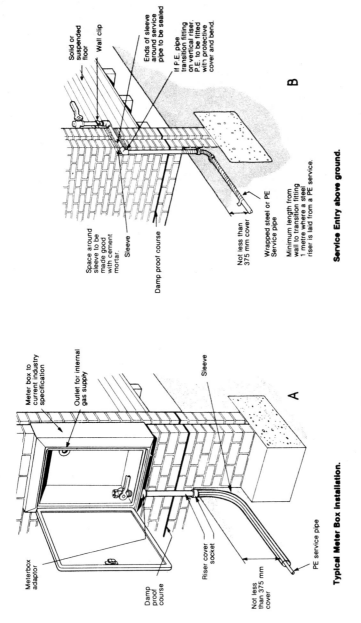

Fig. 15.2 Gas service and meter positions

158 | Construction Technology

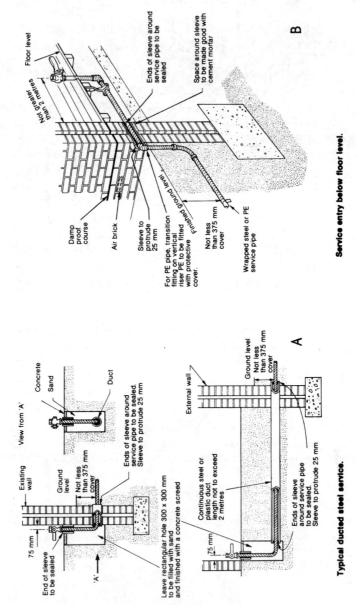

Fig. 15.3 Gas service below ground

the point of entry. Here it is connected to the meter just within the external wall or to the meter box (Figs. 15.2, 15.3, and 15.4A). The responsibility for laying lies with the gas company, who undertake the complete installation if requested.

Installation for natural gas may still be in wrought iron or mild steel for larger diameters or copper or steel for smaller sizes. Polythene pipes and fittings for gas-distribution services, however, are now in wide use. The pipe, which is canary yellow in colour, is available in nominal bore sizes from 12 to 150 mm. Installation requirements in the substructure are shown in Fig. 15.3.

ELECTRICITY. In urban areas electric cables are usually brought underground to an entry-point at ground level for smaller buildings. The service cable terminates at a distribution board which is fitted with a sealing box to prevent moisture from entering the installation (Fig. 15.4B). In country areas the supply is often from overhead cables which are carried to the building at eaves or roof level, then run down the external wall.

The intake position in small installations should be either the garage or near the house porch. Here the consumer's supply control unit is inserted into the wall enabling the meter to be read without entering the house (Fig. 15.4B). The incoming cable is placed in a sleeve in the cavity as shown. It must be possible to bring it in without difficulty and the box positioned to provide an easy route for outgoing cables.

PHONES. The position and arrangement for the entry of the telephone cable should be agreed with the GPO engineers before construction commences so that all wiring can be installed at the carcass stage. The intake can be underground or overhead (Fig. 15.5). In town building the GPO cable could be in the pavement, in which case a sleeve through the external wall would be necessary. For smaller buildings or terraced houses a single pitch-fibre, plastic, or VC pipe of small bore could be brought to the external wall with a lead-in pipe to a connection box as shown. Overhead wires are usually connected to the eaves and then brought into the house via an air brick, ventilator, or small sleeve in the wall.

DRAINAGE. Drains passing through walls or under buildings are subject to building regulations, and the main details of provisions are shown in Fig. 15.6. This method ensures that differential settlement does not cause damage by crushing, and to prevent soil from entering the wall space, asbestos cement sheet or similar material is shaped and slotted over the pipe.

A number of firms now make a full proprietary range of drainage fittings from either UPVC (unplasticized polyvinyl chloride) or VC (vitrified clay). Terminal fittings to waste, soil, rain-water, and w.c.s are supplied complete with adaptors, which make connections through the substructure quite straight-

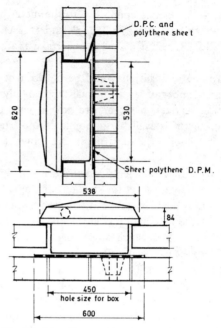

Fig. 15.4A Gas meter box size

forward. Figure 15.6 shows how such connections can be both watertight and flexible.

15.3 Performance characteristics of basic materials

Performance specifications relating to materials and workmanship concerned with service equipment are covered by a number of organizations. The main government bodies are the Building Research Establishment and the British Standards Institution. Performance and standards of building materials and construction are also laid down in B.S. Codes of Practice. These statutory or accepted specifications are normally quoted by reference number in the contract and are binding on the contractor.

Industrial organizations are also concerned with research and development. In recent years, with the production of new and unfamiliar products, independent assessment of materials, components, and processes has become necessary or desirable. Consequently, a European Union of Agrément came into being for the purpose of testing and certifying non-traditional products and techniques and an Agrément Board was set up, with the UK as a member. Thus standardization of quality and dimensions of many materials and products has been brought about. Almost all building materials used in Britain today, however, are covered by British Standards.

Principal service installations | 161

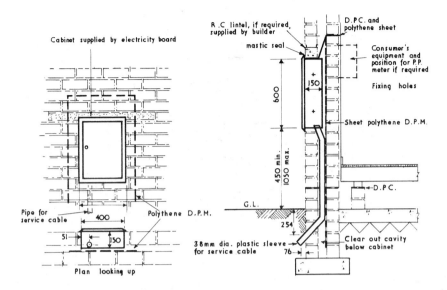

Fig. 15.4B. Electric service entry – domestic

The main properties and performance requirements relating to basic materials connected with water, gas, electricity, telephone, and drainage are: density, strength, electrical and chemical properties, texture, thermal properties and deformation. Failure or weakness in one or more of these characteristics is often due to either *movement* or *deterioration* or both.

WATER. The chief qualities of materials concerned with water supply are strength to withstand pressure and deformation, density to resist leakage, facility to accept joints and fittings, and ability to resist chemical reaction. In the latter case corrosion can occur where certain soils come into contact with metals, cinders, or builders' rubble, with steel or copper, or saline soils with aluminium or galvanized steel. Dissimilar metals such as copper and zinc galvanizing should not come in contact; also rust in base-metals must be guarded against. Metals may also fail through movement caused by expansion, fatigue, vibration, or faulty support, or by abrasion, impact, or fire. Deterioration of metals usually occurs quickly in moist conditions, where condensation is present or where they are exposed to atmospheric pollution.

GAS. Natural gas services, though not required to withstand the pressure of water-pipes or to resist chemical reaction caused by solvents, are nevertheless liable to deterioration; for this reason plastic is now being used. Approved quality plastic gas-pipes are readily recognizable by their canary-yellow colour.

162 | *Construction Technology*

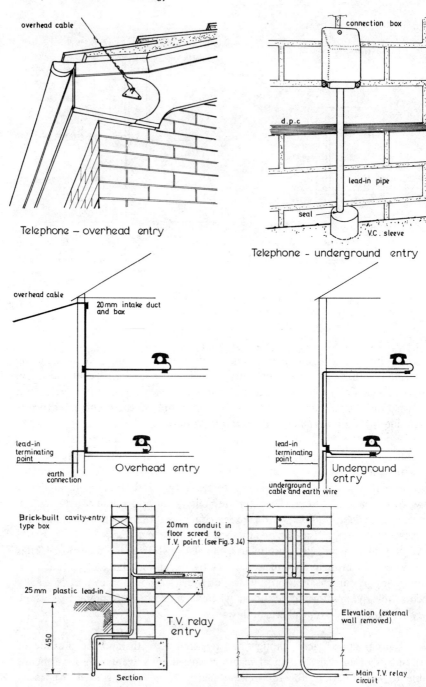

Fig.15.5 Phone and T.V. relay entry

They are easily jointed, but precautions should be taken against puncture, sharp bends, and external pressure.

ELECTRICITY. The main qualities required of electrical materials are conductivity, thermal capacity, insularity, fire resistance, and safety control, depending on their function. All basic materials must be able to withstand the degree of heat for which they have been designed.

Electric cables consist of *conductors* covered by insulation. The cables are normally made of copper, which is one of the best conductors of electricity. They are sized or *rated* to carry the largest amount of current permitted without overheating, and this specific rating must not be exceeded. Cable insulation, whether rubber or PVC, must be adequate for its purpose, but PVC is now the more popular as it can be made in a wide variety of colours and is generally less affected by damp. Cables must be clearly distinguishable by colour to ensure correct wiring of circuits.

The methods of wiring in normal use are:

Sheathed, i.e. separate insulated conductors enclosed in an outer sheath.

Conduit, i.e. single covered insulated cables enclosed in a steel or plastic tube.

Mineral-insulated copper-covered cable, i.e. single strand of copper enclosed in a mineral insulator and the whole encased in a thin copper sheath.

Minimum standards of electrical practice to prevent the risk of fire and shock are laid down in *Regulations for the electrical equipment of buildings* published by the Institution of Electrical Engineers.

TELEPHONES. Phones and wiring operate at 50 volts (as against 240 V. normally used for electrical circuits) and are usually fixed by the GPO. The circuit is almost always powered by batteries, which are automatically recharged from the mains electricity supply. Telephone circuitry is always kept separate from main voltage wiring. The GPO have a booklet prepared in conjunction with the RIBA called *Facilities for telephones in new buildings* which suggests methods and arrangements of installations.

DRAINAGE. There is now a wide choice of materials used in drainage-pipes and fittings. The best known are VC (vitrified clay), either glazed or unglazed, — formerly known as salt-glazed ware or stoneware — cast-iron, concrete, asbestos cement, pitch-fibre, and plastics. Since these are not all equally suitable for all situations, consideration must be given to the particular conditions of the site, such as corrosive soils, soil stability, strength, support, jointing, ground-cover, and type of effluent to be carried.

Drains must be watertight and durable. They must not be affected by the liquids they carry; they must be able to withstand earth movement whether due to underground growths, building-settlement, or soil pressure.

164 | Construction Technology

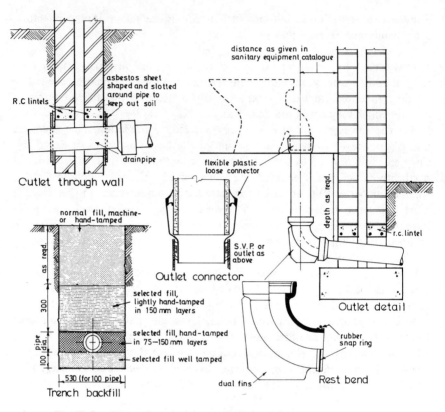

Fig. 15.6 Domestic drainage. Outlet and protection

Concrete pipes and cement joints can be affected by acids, sulphates in the effluent, and by aggressive soils. They are also inflexible.

Pitch-fibre pipes have a good overall chemical resistance to acids and alkalis and to sulphur attack but can be affected by trade wastes containing oils or organic solvents.

PVC and *plastic-wrapped* pipes are generally quite effective but should not be used where hot effluents are likely to be discharged continuously.

Iron pipe is subject to corrosion when in contact with acid or alkali and must be protected by a proprietary bitumen coating.

15.4 Protection of installations

Practically all services require protection of some kind, both of material used and of the installation as a whole. The former is usually applied during manufacture and the latter during or after the work is carried out. Such protection can be against abrasion or impact, acids or chemical action, corrosion, fire, frost, sunlight, vibration, thermal action, or movement. Protection during or after installation only will be dealt with here.

WATER. In most cold water storage and distributions systems some form of protection is necessary. Safeguards against chemical action have already been considered. Tanks and cisterns should always be cleared of sand and grit before being installed to prevent clogging of ball-cocks or valves. Outlets are usually placed a short distance up from the bottom of the tank for this reason but this does not prevent churning. Lids should be provided to reduce risk of frost and to keep the water clean (Fig. 15.1).

Service pipes should be laid 0.76 m below ground as a protection against damage and frost. At the time of laying the foundation, it is usual for the builder to place drain-pipes inside the building rising away from the external wall and through the ground floor slab to act as a protective sleeve (Fig. 15.1). Water supply pipes should not be built into external walls and should be lagged where there is any danger of freezing.

As lagging of new roof spaces is now compulsory, water installations do not suffer from frost to any great extent. Nevertheless, all pipes and cisterns in the roof space should be wrapped except the underside of the cold-water tanks; here the heat rising from the ceiling below is usually sufficient to keep the tank bottom above freezing-point.

As a further protection against damage, the overflow warning pipe from the tank should be carried to the eaves to discharge where it can be seen; also the tank, if of plastic, should be placed on a boarded platform, not simply on joists.

GAS. The service-pipe connecting the consumer's meter with the main supply should run in a straight line as far as possible and be laid to an even fall. The joint to the main is made with a connector which allows the supply to be cut off if no longer required (Fig. 15.2). The trench for the service pipe should be excavated so that the pipe lies directly on the bottom of the trench. Plastic pipes, where permitted, however, may be laid on a bed of sand.

If it is necessary to use a common trench for more than one service, the undertakings concerned should be consulted. It is preferable to avoid running gas and electricity services together because of the danger of corrosion; and in any case no two services of any kind should be in contact with each other. The amount of protection against corrosion for pipes in trenches depends on the nature of the soil. Steel and wrought-iron pipes are normally protected by hessian or bituminous wrapping and covered with asphalt or other form of waterproofing. Pipes above ground are usually protected by a coating of red lead or bituminous paint.

All service-pipes installed by the builder from the meter position to gas points should be either placed in duct or channels in the concrete floor or cast into the screed, and must be suitably protected. Pipes can also be clipped to floor joists when they should be supported without sagging by incombustible materials (Fig. 15.2). Sleeves should be provided for gas-pipes passing through walls as shown. Gas-pipes should not be laid near sources of heat or touch other services particularly electric cables.

ELECTRICITY. The intake location should be as warm and dry as possible, and must be separate from the gas meter and free of condensation. All circuits must be protected from excess current flow by use of fuses or circuit breakers and cable must be of the correct rating for its purpose. Earth connections to installations must be carried out in accordance with I.E.E. regulations to protect the user against electric shock.

No socket outlets are permitted in bathrooms, and switches should be of the pull-cord type; alternatively, ordinary switches can be mounted outside the door. Cables should be kept away from flues and water-pipes (except earth leads). When securing cables to floor joists, buckle clips should be used. Holes through joists for cables are better than notches, but they should not be less than 50 mm from the top. Where notches are unavoidable, they should be only just deep enough to allow a 6 mm clearance of the cable under the floorboarding. Cables should never be bent sharply and when placed in or under solid floors they should be in conduit. Instructions on protection of large installations can be found in I.E.E. regulations.

PHONES. Telephone and other telecommunication circuits should be separated and insulated from electric light and power systems and should not be installed in damp conditions. External cable should not be tacked to lead flashings, wood fascias or frames if possible, but led in as shown in Fig. 15.5. Internal wiring should be ducted if possible, though in small installations this is not always done. Wiring must not be buried below plastering or other finishes, except earth wire where necessary; where dry linings are used, phone or similar wire can be secured to the wall behind these. Phone or bell wire should not be tacked vertically to skirting-board or walls where it can be damaged by vacuum cleaners, polishers, or furniture.

Where it is necessary for wiring to cross flooring it should be protected by flat metal ducting. Sometimes it is more convenient to drop leads vertically from the ceiling. Where wiring is to be installed inside proprietary hollow office partitions, dismantling these for re-positioning later can lead to some inconvenience. Earthing of phone cables is carried out as shown in Fig. 15.5.

DRAINAGE. Protection against fracture in modern drainage installations is usually provided by use of flexible joints to permit freedom of movement. Rubber 'O' rings or flexible plastic sleeve connections are to be preferred to the traditional cement and sand joint, especially where the stability of the ground is suspect. Drains should be laid on a 100 mm underbed of granular material and surrounded by a similar material. This should extend to a height of 300 mm above the pipe if possible, after which the pipe trench should be back-filled to ground level. Should the ground be very soft or where the trench bottom is irregular, the granular underbed may have to be increased in thickness. Details of recommended drain beddings are given in Fig. 15.6. A detail of an expansion joint is also given. The minimum cover over drains should be one metre under fields or gardens and 1.3 metres under roads: shallower drains than this may require a protective reinforced concrete slab over the pipe but not in contact with it.

Where the line of the drain runs alongside a building footing, building regulations require that the pipe be fully surrounded by concrete with appropriate expansion joints at every length. From what has been described, it will be seen that pipe strengths, loads from the ground, and widths of trenches are all related. Figure 3.14 also shows a plan of a simple terrace house with the position of the main services and their entry-points clearly marked. Such a drawing is necessary, particularly where repetitive housing is being erected.

16 | Drainage installations

16.1 Main types of effluent

The primary object of a drainage installation is to carry away foul, waste, and surface water. Methods of disposal vary, but in towns it is usually done by discharging into public sewers, which are maintained by local authorities and are normally sited below roads. Permission to connect from a building has to be obtained, and the work is almost always carried out by the authority itself. Where public sewers are not conveniently placed, other means of disposal can be adopted.

Methods of waste and soil disposal were developed originally by trial and error over a long period of time, and were incorporated into drainage by-laws to govern pipe sizes and layouts of sanitary installations. In comparatively recent times, however, new methods were evolved, based on performance rather than on empirical rules. In this chapter only basic principles will be considered.

Normally, sanitary systems above ground (but including basements) are described as *soil* or *waste* and those below ground as *drainage*. *Soil* is the term used to describe excrement and urine, and *waste* refers to flows from baths, sinks, and lavatory basins. It should be noted that a *lavatory* is strictly a room or fixture for washing as distinct from a *w.c.,* which deals with soil. Both soil and waste are normally discharged together into foul drains and from thence into *foul sewers.* Foul waste also includes trade or industrial discharge for which special arrangements sometimes have to be made. These usually deal with petrol, oil, grease, inflammable liquids, and dangerous chemicals.

16.2 Collection and removal of surface-water

Surface-water drainage is normally concerned with collection and removal from roofs and paved areas. It is largely innocuous except where contamination occurs through impurities caused by industrial pollution or by contact with some waterproof surfaces, such as copper, in certain areas. In order to ease problems of treatment and disposal of effluent, it is normal to isolate foul and waste from surface water by discharging them into *separate* drains, though this is not always the case. Combined and separate systems of drainage will be discussed at Level 2.

The roof shape will often determine the choice of building materials. Roof coverings generally are required to be of pleasing design and appearance, resistant to snow, ice, and dust, and possessing reasonable thermal properties; but their main function is to prevent the entry of rain-water. Coverings for pitched roofs are frequently made of overlapping units or components such as slates, tiles, or sheet materials of corrugated metal or asbestos or flat sheet such as bitumen, copper, lead, or zinc. For smaller buildings with pitched roofs, tiles, or corrugated sheet are popular. Waterproofing of such roofs against abutments is of particular importance and is normally done by flashings of lead (Fig. 16.3).

Flat roofs, i.e. roofs with less than $10°$ pitch are normally made of impervious coverings. These are designed and constructed not only to prevent rain but often to resist acids, natural and industrial impurities left after evaporation has taken place. Vulnerable areas are often those where 'ponding' has occurred, i.e. shallow pools caused by uneven laying.

The question of run-offs from roofs is very important and rates of discharge are usually worked out before building commences. Rainfall run-off varies throughout the country and also at different times of the year. Flat roofs also have a much slower run-off than pitched. The uneven pattern of rainfall intensity is further affected by the geography of the district, effect of wind on buildings, temperature changes affecting the ground, and the impermeability of roof surfaces. Though calculations on rates of discharge can be made, many local authorities still prefer rule-of-thumb methods for finding sizes of pipes, gutters, and outlets. Provision must also be made in the construction for thermal expansion and other movement and precautions taken against lifting by suction which can be very strong.

Roofs, whether flat or pitched, are usually drained by gravity, with the surface-water flowing to eaves gutters laid to falls, and from thence, by downpipes and gullies, to drains leading to soakaways or public sewers (Fig. 16.2). Drainage of paved areas can be affected by a number of considerations such as high-intensity rainfall, storm frequency, geography of the area, slope of the ground, and season of the year. The impermeability of the paved surface itself and available fall are other factors to be noted when installing drainage. The type of gully may also require consideration, particularly if the area is to be used for washing purposes or where grit or sediment is likely to collect. Two types of flat roof outlet are shown in Fig. 16.1 and the standard gullies for draining paved areas are shown in Fig. 16.2.

Public sewers are normally used in towns for the disposal of surface-water. These accept not only roof waste from downpipes leading to trapped gullies but also the discharge from roads, pavements, and paved areas (Fig. 16.2). The latter are usually laid to shallow falls which terminate at grated gullies from which branch drains lead to public sewers.

Sometimes where public sewers are used for both surface- and foul-water discharge, overloading occurs and it becomes impracticable to accept both types of effluent. In such cases soakaways must be used for surface-water (Fig. 16.4). These are normally pits dug within the tenant's own boundary and almost filled with rubble or hardcore, after which the hole is back-filled up to

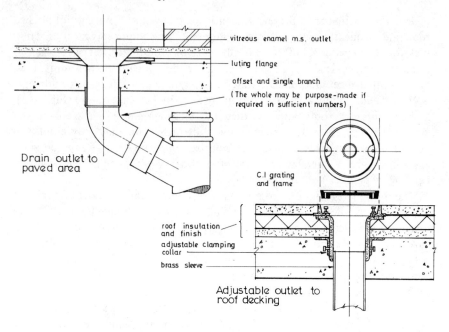

Fig. 16.1 Drainage of roof and paved areas

ground level. Sometimes a concrete base and perforated walls are provided in accordance with the requirements of the local authority (Fig. 5.9).

In areas liable to flooding or where the subsoil water or water-table is high, it may be necessary to insert land drains in the ground, laid to falls depending on the gradient. These pipes are porous and laid with butt joints in narrow trenches. The spacing and depth will depend on the nature of the ground and its permeability. This can be tested by digging trial holes and noting the water seepage. The outflow from land drains normally discharges into ditches.

16.3 Rain-water eaves-gutters, pipes, and fittings

Rain-water goods for eaves-gutters, and downpipes are normally made of asbestos cement, galvanized pressed steel, cast iron, aluminium, copper, zinc, and plastics. Though pressed steel and aluminium are now in general use, plastics appear to have captured the domestic market, and are widely used in less expensive types of building. Cast iron, once in almost sole use, is not often seen in new building. External rain-water goods in use for relatively small roof areas are covered by the following British Standards:

Asbestos cement	B.S. 569
Pressed steel, galvanized,	B.S. 1091
Cast iron	B.S. 640

Drainage installations | 171

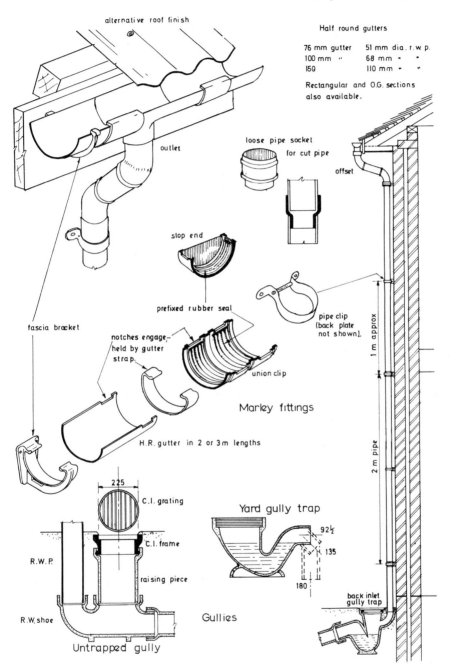

Fig.16.2 Eaves gutters and rainwater fittings

Aluminium   B.S. 2997
Wrought copper and zinc   B.S. 1431
UPVC rain-water goods   B.S. 4576

Eaves-gutters are usually made to standard sizes and designs, i.e. half-round (H.R.), ogee (O.G.), or moulded in section. Plastic gutters are also made in rectangular or box section. Larger sizes of gutters are not standard but usually purpose-made for large areas of roof, but only standard sizes will be considered here. Sizes for H.R. gutters range from 75 to 200 mm diameter but 75 to 100 mm is normal for smaller roofs.

Rain-water pipes are normally made from the same material as the gutter itself, and manufacturers usually supply the whole system complete with fittings. Downpipes of 50 to 100 mm diameter are selected to match gutters for small to medium installations. Older systems are often replaced wholly or in part by plastics, particularly where appearance of different materials is of little importance. PVC pipe is used for plastics systems though polythene downpipe is also used as it is stronger. It is usually black in colour to make it more resistant to daylight.

FITTINGS. Most fittings vary according to the material of which the system is made. Eaves gutters with fascias include unions, outlets, stop ends, internal and external angles, and brackets which may be shaped to suit either rafters or fascias. Sometimes fittings are moulded or cast on as an integral part of the gutter or pipe, but are often supplied as accessories. The latter also include offsets, bends, branches, shoes, and rain-water heads, and Fig. 16.2 shows some of these. Balloon guards of wire or plastic are often fitted to gutter outlets where leaves or debris may cause blockage.

JOINTING. Methods of jointing vary with the material used. It may be necessary only to provide a durable, watertight connection, but with certain materials allowance must be made for expansion. Generally, *all* joints should be lapped in the direction of the flow.

Asbestos cement fittings are normally jointed with either a mastic jointing compound or special synthetic rubber, held together tightly with screws. Spigot and socket downpipe joints are made loose and there is no dimension provision for jointing material; this permits thermal movement to take place.

Galvanized pressed steel of light gauge is jointed in the same way as asbestos cement above, but using either red lead putty or cementitious cold caulking compound. Downpipe joints are left loose as for asbestos cement pipes.

Cast iron fittings are jointed in a manner similar to that of pressed steel.

Aluminium rain-water systems should be uniformly of one kind of metal and should not come into contact with water running from copper surfaces. Jointing is done with any mastic-based compound or bitumen butyl rubber. Compounds containing copper or lead should not be used. Downpipes may be

Drainage installations | 173

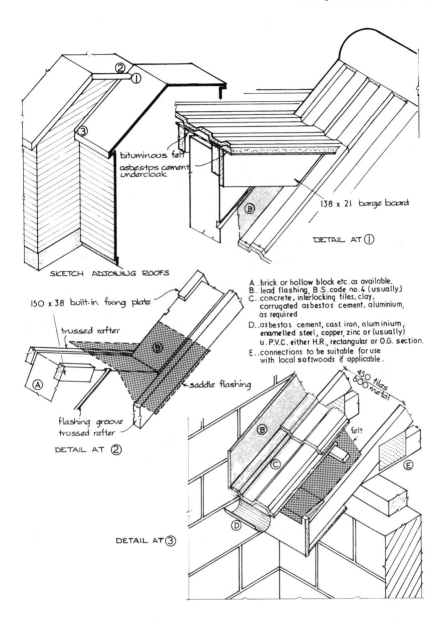

Fig. 16. 3. Stepped and staggered roofs

loose or as described for gutters and sockets can be tapered for caulking if required.

Copper and zinc. Jointing is usually done by lapping 37 mm in the direction of the flow. Soldering should be avoided if possible, but where necessary should be done with care. Expansion gutter joints must be allowed for every 15 m. Downpipes are made so that the socket end is slightly larger than the spigot to permit a telescoped joint 50 mm long.

Plastics. Each manufacturer has developed his own system of fixing and jointing. A number have snap-on joints with composition built-in rings to permit both expansion and watertightness (Fig. 16.2). New designs are still being produced, some of which combine soffit, fascia, and gutter in one piece which simply clips into place. Generally, however, jointing of both gutter and downpipe is done by means of foamed neoprene or rubber gaskets (Fig. 16.2).

FIXING. Methods vary greatly but the main difference lies in the need for flexibility and whether fittings are integral or supplied loose as accessories.

Asbestos cement. Gutters are usually fixed by means of galvanized brackets screwed to the fascia board at not more than one-metre centres. Downpipes are made uneared and fixed by means of galvanized m.s. ring clips fitted below each socket and designed to give a 37 mm offset from the wall. Loose pipe sockets are also supplied (Fig. 16.2).

Pressed steel gutters are fixed at joints with small bolts with slotted mushroom heads. Intermediate holes are also provided for fixing to feet of rafters or fascia with wood screws.

Cast iron is fixed as for asbestos cement.

Aluminium gutters are fixed every 2 metres into either aluminium or galvanized steel brackets screwed to the fascia. All bolts and other fixings must be of aluminium. Downpipes usually have flat ears to take galvanized pipe nails driven into hardwood plugs in the wall.

Copper and zinc gutters are supported at every 750 mm, with the point of support coinciding with alternate stays. Brackets are of hard copper and gutters are fixed by screwing. Downpipes are fixed by means of wood screws inserted through the ears into hardwood plugs designed to keep the pipe clear of the wall. Pipe screws should be of copper alloy or galvanized steel for zinc pipes.

Plastic rain-water goods usually have their own proprietary methods of fixing which may be seen in manufacturers' catalogues. Typical examples are shown in Fig. 16.2 and 15.6.

16.4 Underground drainage systems: general principles

The general principles to be observed in piped systems of underground drainage are:

Drainage installations | 175

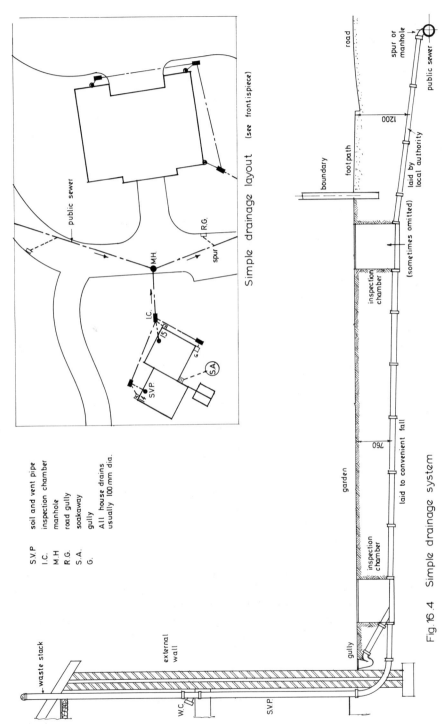

S.V.P soil and vent pipe
I.C.   inspection chamber
M.H   manhole
R.G.   road gully
S.A.   soakaway
G.   gully
       All house drains usually 100 mm dia.

Simple drainage layout (see frontispiece)

Fig. 16.4 Simple drainage system

They must be able to accept immediately the flow from any sanitary appliance connected to the system;

The flow must be discharged completely and without delay to the sewer or treatment plant;

No water must emerge from the system under any conditions of flow; and

Air must at no time be allowed to escape from the pipes into the building. Solids in the waste must be carried by the flow without becoming lodged or cause blockage.

To conform with these requirements, drains must be:

Watertight, which is done by selection of suitable material and jointing systems and by satisfactory execution.

Free-flowing, by using pipes with smooth interior surfaces, properly aligned joints and of adequate minimum diameter. No bottlenecks or restrictions in the pipe diameter must occur.

Laid to correct falls and to an even and regular gradient.

Air-sealed by traps at all entries to the system in the case of soil-water systems, and with the installation designed to accommodate air pressure fluctuation which could unseal traps.

Durable against deterioration due to chemical action of liquid waste or from the nature of the external soil, or failure of the jointing materials.

Resistant to failure against movement due to settlement or inadequate ground cover, or poor bedding support, or root growth, or lack of flexibility in jointing. (Materials for drains were considered briefly in the last chapter.)

Accessible for inspection, cleaning, and repair.

Economical in design and planning. Systems should incorporate good siting and layout, and grouping of appliances to the best advantage.

Ideally it is desirable that the building should be sited so that the drainage can be planned to discharge by gravity to the sewage pipe or other disposal system without excess fall or without pumping or other mechanical assistance. Clearly, the fewer the number of entries to the system the more economical it becomes and the shorter the drain lengths the better. Long lengths of drains usually involve deep excavations and possible pumping, and also deeper access points such as manholes and inspection chambers.

Should the local authority be required to install a sewage system to accommodate a new housing estate or industrial complex, early consultation between architect and engineer could result in the provision of spurs or connecting points for branch drains made at a convenient depth to meet the required gradients of the drainage layout (Fig. 16.4). It is usually economic to discharge into the local sewer and this can be insisted upon if it is within range.

Where excessive surface water is likely to cause overloading in the sewer it may be necessary to dispose of this into a separate system. This is normal practice but where the annual rainfall is fairly low, as in London for example, surface water and foul drains may be combined. This has certain advantages and disadvantages which will be considered at Level 2.

16.5 A typical drainage system

A typical house drainage system showing the general principles of layout in section together with a simple layout plan is given in Fig. 16.4. In this example all surface-water is drained into soakaways, a method which is now fairly common. Road drainage and house waste, however, has been taken to the foul sewer. The inspection chamber and manholes consist of brick or concrete pits usually at the intersection points where drains change direction. This enables the drain to be cleared of blockage and also to be rodded through if necessary. Construction of chambers and soakaways will be dealt with at Level 2.

Guide to further reading

Books

Mitchell's *Building Construction,* Metric editions: Batsford, London:
   Burberry, P. (1975) *Environment and services.*
   Everett, A. (1970) *Materials.*
   Stroud Foster, J. (1973) *Structure and fabric, Part 1.*
   Stroud Foster, J. and Harington, R. (1976) *Structure and fabric, Part 2.*
   King, H. and Everett, A. (1971) *Components and finishes.*
Reid, D. A. (1973) *Construction principles, 1: function.* George Godwin, London.
Scott, J. S. (1974) *Dictionary of building.* Penguin, Harmondsworth.
Specification (Annual), Architectural Press, London.

Book-lists

A comprehensive catalogue of books for the construction industry as a whole is published annually by the *Building bookshop.*
An annual *RIBA book-list* is also available to architects and students.

Sources of information

The Building Research Establishment (B.R.E.) produces an *Information Directory* annually, which includes a complete list of current government publications on building activities and requirements, materials, elements, and the built environment.
    The Department of the Environment (D.O.E.) also publishes a leaflet annually called *How to find out,* which gives details and sources of information selected from 550 organizations in the construction industry. All the above lists and catalogues may be obtained direct from The Building Centre, 26 Store Street, London WC1 7BT.

Glossary

Admixture: A material other than coarse or fine aggregate, cement, or water, added in small quantities during the mixing of the concrete to produce some desired modification in one or more of its properties (B.S. 2787: 1956).
Algae: Black seaweed growth on roofs and walls.
Apron lining: A board used to form a finish at the edge of the floor round a stair-well or other similar opening (B.S. 565: 1972).

Bitumastic compound: A water sealer containing bitumen which remains pliable enough to accommodate movement in joints.
Bond: (1) An interlocking arrangement of structural units within a wall to ensure stability. (2) Adhesion between materials in composite construction (B.S. 3589: 1963).
Breeze: Usually understood to mean clinker but has sometimes been used to refer to coke-breeze (B.S. 2787).

Casement door: A hinged door or pair of doors almost wholly glazed (B.S. 565: 1972).
Casement window: A window in which one or more lights are hinged to open (B.S. 565: 1972).
Carcass: Framing in position before addition of covering (B.S. 565).
Cladding: The external non-loadbearing covering to the frame of a building or structure.
Code of practice: Good building practice recommended by the B.R.E.
Column: An upright (vertical or near-vertical) loadbearing member whose length on plan is not more than four times its width (B.S. 3589).
Component: A section unit or compound unit (B.S. 3589).
Cross-wall construction: A type of construction in which floor and roof loads are carried entirely on walls running across a building (B.S. 3589).
Curtain walling: A non-loadbearing wall constructed outside and continuously over a structural frame to enclose a building or structure (B.S. 3589).

Damp-proof course: A layer or sheet of material placed within a wall, column, or similar construction to prevent the passage of moisture (B.S. 3589). (A damp-proof membrane is a damp-proof course within a floor or flat roof.)
Dead load: The load represented solely by the weight of walls, partitions, roofs, floors, and other permanent constructions including finishings (B.S. 3589).
Dimensional co-ordination: Agreement made between the manufacturers of building units and the designers in order to simplify assembly by standardizing sizes.
Dry rot: A type of decay of timber in buildings caused, in the United Kingdom, by the true dry rot fungus, *Merulius lacrymans* (B.S. 565).

Eaves: The lower edge of a roof at the face of the wall (B.S. 565).

Edge beam: A stiffening beam at the edge of a slab (B.S. 2787).
Element: Part of a building or structure having its own functional identity. Examples are *foundation, floor, roof, wall* (B.S. 3589).

Fascia: A long and relatively narrow upright face at the eaves or cornice or over a shop front.
Finishing: Fixtures to and treatment of surfaces to convert the carcass into a complete building excluding services (B.S. 3589).
Flashing: A strip of impermeable sheet material fixed so as to protect a joint or surface from the entry of rain-water (B.S. 3589).
Flat roof: A roof the pitch of which is 10° or less to the horizontal (B.S. 3589).
Float: A flat-faced wood (or metal) hand tool for spreading or smoothing concrete or mortar: *verb;* to use a float (B.S. 2787).
Floor: A construction that provides the surface on which one walks in a building or structure (B.S. 3589).
Floor slab: A slab forming the continuous loadbearing structure of a floor and spanning between supports or laid on the ground (B.S. 3589).
Formwork: (shuttering). A temporary construction to contain wet concrete in the required shape whilst it is being cast *in situ* (B.S. 565).
Foundation: A construction to spread loads applied to the supporting soil (B.S. 3589).
Framing: Timber used in the structural work of a building (B.S. 565).

Glazed door: see casement door.
Glazing: (1) Glass or other material used in filling in an opening to admit light and exclude weather (B.S. 3589). (2) The art of fixing glass in a frame.
Granolithic finish: A surface layer of granolithic concrete which may be laid on a base of either fresh, green, or hardened concrete.
Grid: A rectangular network of lines used in planning or setting out (B.S. 3589).
Ground beam: A horizontal beam of (usually) r.c. built at ground level and supported by piles or pads.

Half-round: Of semicircular section.
Hard standing: A paved area.
Head: (1) A horizontal member supported by posts.
   (2) A transverse member forming top of an opening.
   (3) The top member of a frame usually horizontal (B.S. 565).
Herring-bone strutting: Floor strutting between joists to stiffen them (B.S. 565).
Humidity: Moisture content in the atmosphere. (This is normally measured by a hygrometer.)

Impact load: An imposed load whose effect is increased due to its sudden application (B.S. 3589).
Industrial building: A building used principally for winning, assembling, processing or treating materials for the purpose of manufacture or trade.
In situ: Literally 'in place', referring to material or components that are cast or assembled in their permanent position in a building or structure as distinct from being cast or assembled before installation (B.S. 3589).

Jamb: A vertical side member of a door frame or door lining or window-frame (B.S. 565).

Joist: One of a number of members spanning horizontally between supports to carry flooring or ceiling or both.

Kerb: A low barrier rising above the surrounding surface.
Key: Irregularity or indentation of a surface which creates a mechanical bond for plaster.
Knock-down: (k.d.) To dismantle into separate components for ease of packing and transport.

Load: A force acting on a structure or a member (B.S. 3589): *imposed load:* The load on a structure arising from its position and function other than a dead load.
Loadbearing wall: A wall designed to support a load in addition to its own weight and wind pressure on its surface.

Made ground: Ground built up with excavated material or refuse as distinct from the natural, undisturbed soil (B.S. 3589).
Masonry: Construction of stones, bricks, or blocks (B.S. 3589).
Module: A common unit particularly specified for dimensional co-ordination.
Moisture content: The amount of moisture in timber or other material expressed as a percentage of its oven-dry weight (B.S. 565).
Mortice and tenon joint: A joint in which a tenon on the end of one member is fitted into a mortice cut in the other member.
Mullion: An intermediate vertical member in a window frame, door frame or similar structure.

Neoprene: Synthetic rubber-like plastic which resists heat and light.

Panel: (1) A distinct portion of a floor, roof slab or wall supported by a frame (B.S. 3589).
(2) A filling to a space surrounded by framing (B.S. 565).
(3) An area of concrete between beams, columns or other boundary lines (B.S. 2787).
Partition: (1) A wall whose primary function is to divide space within a building or structure (B.S. 3589).
(2) A continuously supporting frame and facings or infilling. (B.S. 3589).
Party wall: A wall common to two buildings or two pieces of land.
Pile: A support driven into or cast *in situ* in the ground.
Pitch: The angle of a slope to the horizontal. (B.S. 3589).
Pitch-fibre pipe: Pipe of wood-fibre pulp impregnated under vacuum with coal tar pitch to about 75 per cent of its total weight.
Plywood: A product made up of plies and adhesives in which the plies are crossed to improve the strength properties.
Portland cement: A finely ground powder made in accordance with B.S. 12 which when mixed correctly with water hardens and adheres to suitable aggregates.
Prefabrication: Fabrication on or off the site before incorporation into a building or structure (B.S. 3589).

Raft foundation: A supporting slab continuous over the whole area covered by a building or structure (B.S. 3589).
Rafter: One of a number of inclined members to which a roof covering is fixed.

Rendering: The application by means of a trowel or float, of a coat of mortar.
Retaining wall: A construction providing lateral support to the ground or a mass of other material (B.S. 3589).
Reveal: The vertical face revealed in the thickness of an opening or the depth of a recess. (cf. *soffit*).
Ridge board: The longitudinal member at the apex of a roof (B.S. 565).

Screed: (1) A layer of mortar on a hard surface (also as verb, to *screed*).
(2) A strip, usually of wood or metal, used as a guide for striking off or finishing a surface.
(3) A strip moved over a guide to strike off or finish a surface.
Services: Installations for (1) the introduction into and distribution within a building or structure of water, air, gas, liquid fuel, electricity, heat or other source of energy; (2) the disposal of waste from a building or structure; or (3) fire-fighting within a building or structure. The term does not apply to lifts, escalators, or similar mechanical equipment (B.S. 3589).
Shuttering: See *formwork*.
Sill: (1) A construction extending over a wall below an opening designed to throw rain-water clear (B.S. 3589). (2) the horizontal bottom member of a window frame, door-frame, or other frame or framing (B.S. 565).
Skirting: A finishing member fixed to a wall or other vertical surface where it meets the floor.
Sleeper wall: (also *dwarf* or *honeycomb*) A low wall supporting the joists of the lowest floor of a building (B.S. 3589).
Soffit: An exposed under-surface including that of a ceiling (B.S. 3589).
Solid floor: A floor laid direct on the ground or over a continuous filling.
Stairs: See Chapters 11 and 14.
Stile: A framed outer vertical member of a door or sash.
Storey: The space between two floors or between a floor and roof.
Structure: An organized combination of connected units constructed to perform a function or functions requiring some measure of rigidity.
Substructure: The part of a building or structure below the level of the adjoining ground.
Superstructure: The part of a building or structure above the level of the adjoining ground.
Suspended floor: A floor that spans between supports.

Throating: A groove under a sill, coping, or moulding.
Transom: An intermediate horizontal member of a window-frame, door-frame or similar structure.
Trimming: Trimmers and trimming joists or trimming rafters forming an opening.
Tusk tenon: A combined tenon and housed joint for bearing timbers.

Unit: Building material which is formed as a simple article complete in itself but which is intended to be part of a compound unit or building or structure. Examples are brick, block, tile, lintel (B.S. 3589).

Wall hanger: A metal bracket suspended from a wall to support a member.
Wall plate: A member built into or on a wall to distribute the load from a timber floor or roof.
Water bar: A metal strip in the sill of a door or window designed to prevent penetration of water.

Weather moulding: a moulded projecting member fixed to the bottom rail of an external door or sash to divert water from the sill or threshold. (A *weather board* is similar but unmoulded.)

Wind load: The force exerted by wind on a structure or part of a structure (B.S. 3589).

Index

Acid in plaster, 77
algae, 81, 179
apron lining, 149, 179

Basements, 46
beds and pavements:
  beds, 55
  brick pavers, 58
  clay tiles, 58
  hard surfaces, 58, 134, 180
  hoggin, 58
  kerbs, 42, 57, 181
  pavements, 55, 57, 169
  pre-cast flags, 58
  tarmacadam, 58
bitumastic compound, 58, 179
bitumen felt, 132
bore- and trial-holes, 42, 46, 170
building function, 34

Capillarity, 80
carcass, 35, 179
cement, *see* Portland cement
cladding, 13, 35, 67, 179
climate: 7
  conditions, 74
  weather exclusion, 80, 90, 104, 183
Code of Practice, 104, 142, 160, 179
communication, 9
components, 7, 8, 16, 69, 71, 113, 179
compound units, 5, 71, 113
compound unit components, 71, 113
contractor, 11
co-ordinating space, 5, 95, 145
cross-referencing, 17

Damp-proofing, 34, 129, 131, 179
designer, 11, 13, 74
dimensional co-ordination, 4, 8, 27, 71, 146, 179
doors:
  appearance, 91
  basic types, 91
  casement, 94, 179
  durability, 91
  feasibility, 91
  fire resistance, 91
  flush, 94, 97
  door sets, 95, 96
  frame assembly, 97
  framed, 92, 93, 94
  glazing, 92, 96, 180
  handing, 102
  ironmongery, 99, 102
  matchboarding, 94, 99
  panelled, 92, 181
  performance requirements, 90
  sizes, 96
  sound and thermal insulation, 91
  steel casement, 94
  stile, 92, 182
  strength, 90
  types, 92, 93
  unframed, 94
  water bar, 98
  weather protection, 90, 98, 183
drawings:
  assembly, 16, 29, 31
  block, 29, 31
  component, 16, 29, 31
  dimensioned, 27, 31
  documentation, 16
  freehand sketching, 19, 22
  graphic symbols, 16, 25
  grid, 16, 27, 31, 180
  isometric, 20
  location, 16, 29, 31
  measured, 17
  notation and symbols, 25, 27
  numbering, 17
  orthographic projection, 19
  production, 27, 29
  scales, 22, 29
  schedules, 16
  titles, 33
drainage:
  foul sewer, 168
  installation, 164, 176
  jointing, 172
  pipes, 163, 181
  pollution, 168

removal of water, 168
roofs, 105, 168, 172
services, 154, 159, 163, 166
soakaway, 42, 47, 169, 177
soil waste, 169
systems, 177
types of effluent, 168
underground, 174

Electricity, *see* Services
elements: 180
  built environment, 3
  construction, 69
  functional, 3, 113
  internal, 113
  location, 3, 7, 8
  primary, 34, 113
  secondary, 35
excavation:
  bottoms, 51
  cut and fill, 44
  made ground, 41, 49, 181
  removal of water, 46
  soil types, 41, 50
  types, 42
external envelope, 3, 63, 73, 81, 105

Factor of safety, 74
fascia, 172, 174, 180
feasibility of project, 79
finishings and finishes: 35, 180
  applied, 37
  bituminous emulsion, 58
  dry, 126, 127
  floated, 139, 180
  floor, 137, 140, 141, 180
  granolithic, 139, 180
  integrated, 35
  key, 124, 126, 181
  paints, 127
  preformed, 36
  rendered, 126, 182
  screeded, 135
  self-finish, 35, 126
  wet, 126, 135
fire resistance, 79, 81
fittings, 18
floors: 180
  concrete, 132
  construction, 115
  damp exclusion, 129
  damp proofing, 129, 139
  dry rot, 134, 179
  edge beams, 17, 46, 52, 55, 132, 180
  fire, 128
  function, 113, 128
  granolithic, 139, 180
  ground, 129
  in situ, 115, 135
  joint hangers, connectors, 137, 182
  jointless, 139
  joists, 129, 134, 135, 181
  loading, 61
  panel, 62, 181
  screed, 135, 139, 182
  sheet, 137, 139
  skirting, 148, 149, 182
  slab, 131, 139, 180
  solid ground, 129, 182
  storey, 114, 182
  strength, 128
  strutting, 137, 180
  suspended, 114, 128, 132, 135, 182
  terrazzo, 140
  thermal insulation, 129
  tiles, 140
  trimming, 135, 139, 182
  wall plate, 130, 132
  wood, 132
formwork, 13, 36, 180
foundations: 180
  beds, 55
  cement concrete, 51
  choice, 48
  clay, 50
  depth, 55
  differential movement, 50
  erosion, 49
  flooding, 49
  function, 34, 48
  ground beam, 54
  hardcore, 52, 131
  infilling, 50
  internal, 121, 122
  loading, 49
  pad, 46, 52
  piling, 46, 48, 181
  primary materials, 51
  raft, 48, 55, 61, 181
  scour, 49
  settlement, 50, 52
  slab, 44, 52
  slip, 50
  slope, 50
  stability, 49
  stepped, 44, 54
  strip, 44, 52, 55, 61
  subsidence, 49
  subsoil movement, 50
  understructure, 34
  unit construction, 51
  walls, 119

water, 46, 50
framing and framework, 69, 71, 118, 180
function of environment, 3, 73

Gas, *see* Services
ground:
    beam, 54, 180
    condition, 49, 55
    floor, 128, 129, 180

Humidity, 76, 180

Industrialization, 11, 121, 131, 180
insulation:
    floor, 131
    sound and noise, 76, 81
    stair, 142
    thermal, 74, 80, 119
in situ work, 4, 8, 11, 13, 35, 48, 52, 58, 69, 113, 115, 126, 135, 140, 180
ironmongery, 89, 99

Joints:
    brickwork, 126
    connectors, 137
    design, 8, 12
    drainage, 164, 167
    hanger, 138
    mortice and tenon, 137, 181
    pipe, 154, 159, 161, 172
    tusk tenon, 129, 182

Knock-down, 71, 181

Lightweight concrete, 76, 124
loads and loading: 49, 180, 181
    dead, 48, 61, 179
    distributed, 63
    impact, 48, 180
    imposed, 48, 61
    live, 48
    loadbearing, 121, 181
    load factor, 74
    non-loadbearing, 63, 65, 121
    point, 63
    superimposed, 48
    wind, 48, 64, 104, 183

Manufacture, 12
materials:
    admixture, 81, 179

basic, 5, 160
concrete, 51, 58
earth, 67
external walls, 80
finishes, 36
protection, 78
roofs, 169
testing, 13
mechanical plant, 13, 42, 44, 52
metal:
    corrosion, 71
    extrusion, 71
    processing, 71
modular co-ordination, 4
modules, 4, 181
moisture content, 77, 181
mortar admixture, 81, 179
movement and moisture, 77

Neoprene, 174, 181

Openings:
    arch, 24
    floors, 135
    head, 85, 95, 120, 180
    lintels, 120, 123, 124, 126
    jamb, 98, 120, 124, 180
    reveal, 120, 182
    sill, 85, 98, 120, 182
    soffit, 120, 182
    walls, 124
orientation, 33

Panels, 62, 71, 92, 114, 115, 181
performance specification, 8, 103, 160
pitch fibre, 159, 164, 181
plasterboard, 118, 121, 127, 135, 141
Portland cement, 51, 77, 181
prefabrication, 5, 8, 13, 114, 181

Quarry tiles, 58, 140
Quilt, 134

Rain-water:
    eaves gutter, h.r., 170, 172, 180
    fittings, 172
    goods, 170
    jointing, 172, 174
    penetration, 74
    pipes, 172
    roof run-off, 169
reinforced concrete, 74

Index | 187

roofs:
   appearance, 105
   basic forms, 107
   drainage, 169
   durability, 105
   eaves, 108, 172, 179
   fascia, 171, 180
   feasibility, 107
   fire, 105
   flashing, 169, 173, 180
   flat, 61, 107, 169, 180
   gutters, 170, 172, 180
   insulation, 76, 104
   performance, 103
   pitch, 104, 108, 151
   rafter, 110, 181
   ridge, 108, 182
   shapes, 103
   stepped, 173
   weather exclusion, 104
   wind pressure, 104

Sections, 5, 71
security, 4, 10, 73
services: 182
   basic requirements, 10, 18, 153
   drainage, 154, 159, 163, 166
   effluent, 168
   electricity, 122, 154, 159, 163, 166
   entry and outlet, 156
   gas, 154, 156, 161, 165
   installations, 153, 182
   performance, 160
   telephone, 122, 154, 159, 163, 166
   protection, 164
   water, 153, 156, 161, 165
shuttering, *see* formwork
site:
   access, 9
   communication, 9
   conditions, 10, 41, 49
   construction, 11, 13
   choice, 9
   layout, 11, 12
   location, 9, 141
   organization, 11
   plan, 41
   plant, 13
   project, 18
   security, 10
   slope, 50
   standardization, 12
   subsoil, 50

slab thickening, 44, 55, 121, 131

stairs: 182
   appearance, 143
   balustrade, 144
   constituent parts, 143
   construction, 145, 146
   feasibility, 143
   fire, 143
   function, 142
   insulation, 142
   pitch, 146, 181
   straight flight, 115, 143
   strength, 142
   reinforced concrete, 115
   types, 115
standardization, 4, 12, 69
storey, 114, 142, 182
structure: 182
   appearance, 79
   basic types, 64
   columns, 64, 179
   concept, 61
   constituent parts, 34
   design, 7, 74
   forces, 67
   forms, 67
   frames, 3, 69
   function, 61
   strength, 67, 69, 73
   strip foundation, 44, 52
   substructure, 18, 34, 41, 61, 156, 182
   superstructure, 18, 34, 61, 182
   types, 65, 67

Telephone, *see* Services
throating, 91, 182
timber and timber-based products:
   chipboard, 139
   floors, 135
   frames, 118
   framed partitions, 121
   moisture content, 77
   movement, 77
   plywood, 135, 139, 181

Ultraviolet light, 78, 84
UPVC, 172
units, 5, 24, 71, 113, 182
universal classification, 17, 18

Ventilation:
   external envelope, 73
   floors, 132
   underfloor, 129
   windows, 84, 86

veranda, 114
vibration, 81

Walls:
    block, 81, 124
    bond and bonding, 84, 124, 179
    breeze, 124, 179
    brick, 120, 122
    cavity, 81, 130
    cross-walls, 61, 64, 67, 121, 179
    curtain, 13, 63, 74, 179
    dry lining, 127
    dwarf, 130
    elements, 113
    external, 7, 80, 119
    fender, 133
    finishes, 126
    fire, 114, 119, 122
    foundation, 121
    function, 118
    hanger, 137, 182
    insulation, 119, 121
    internal, 113, 118, 122
    key, 126, 181
    loadbearing, 63, 118, 121, 122, 181
    masonry, 81, 118, 181
    membranes, 118, 122
    monolithic, 118
    mortar, 8
    non-loadbearing, 118, 119, 121
    partitions, 35, 44, 113, 118 121, 126, 181
    party, 114, 118, 122, 181
    plate, 130, 132, 133, 182
    retaining, 44, 46, 182
    self-finish, 126
    sleeper, 132, 182
    solid, 81
    strength, 80
    ties, 84
    weather exclusion, 80
    weepholes, 131
    wet finish, 126
water:
    bar, 91, 182
    deterioration, 77
    removal, 46, 168
    run-off, 58
    services, 153, 156, 161, 165
wet construction, 7, 11, 69, 126
windows:
    air movement, 84
    casement, 82, 87, 179
    daylight, 84
    design, 86
    double glazing, 80, 84
    EJMA, 86
    identification, 89
    mullion, 85, 181
    noise insulation, 86
    performance, 84
    sizes, 86
    solar radiation, 84
    steel, 87
    timber, 86
    transom, 85, 182
    types, 85, 87
work size, 31

Yale locks, 101

Zinc, 78, 170, 174

Useful conversion factors

	Multiply by
Square yards to square metres	0.836
Square metres to square yards	1.196
lb/yd^2 to kg/m^2	0.5425
kg/m^2 to lb/yd^2	1.8433
in^2/ft to mm^2/m	2116.64
mm^2/m to in^2/ft	0.000472
lb per lin. foot to kg per lin. metre	1.48816
kg per lin. metre to lb per lin. foot	0.67178

Force

lbf to N (Newtons)	4.448
N to lbf	0.225

Stress

lbf/in^2 to N/mm^2	0.0069
N/mm^2 to lbf/in^2	145.04

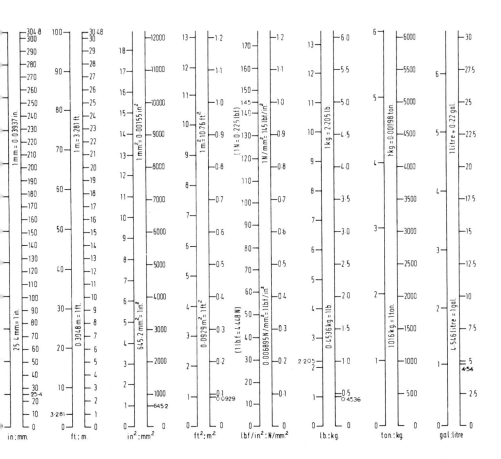